REDUCING THE THREAT OF IMPROVISED EXPLOSIVE DEVICE ATTACKS BY RESTRICTING ACCESS TO EXPLOSIVE PRECURSOR CHEMICALS

Committee on Reducing the Threat of Improvised Explosive Device Attacks by Restricting Access to Chemical Explosive Precursors

Board on Chemical Sciences and Technology

Division of Earth and Life Studies

A Consensus Study Report of

The National Academies of
SCIENCES · ENGINEERING · MEDICINE

THE NATIONAL ACADEMIES PRESS
Washington, DC
www.nap.edu

THE NATIONAL ACADEMIES PRESS   500 Fifth Street, NW   Washington, DC 20001

This activity was supported by Contract No. HHSP233201400020B/HHSP23337050 with the U.S. Department of Homeland Security. Any opinions, findings, conclusions, or recommendations expressed in this publication do not necessarily reflect the views of any organization or agency that provided support for the project.

International Standard Book Number-13:   978-0-309-46407-9
International Standard Book Number-10:   0-309-46407-2
Digital Object Identifier: https://doi.org/10.17226/24862
Library of Congress Control Number: 2018930758

Additional copies of this report are available for sale from the National Academies Press, 500 Fifth Street, NW, Keck 360, Washington, DC 20001; (800) 624-6242 or (202) 334-3313; Internet, http://www.nap.edu.

Copyright 2018 by the National Academy of Sciences. All rights reserved.

Printed in the United States of America

Suggested citation: National Academies of Sciences, Engineering, and Medicine. 2018. *Reducing the Threat of Improvised Explosive Device Attacks by Restricting Access to Explosive Precursor Chemicals.* Washington, DC: The National Academies Press. doi: https://doi.org/10.17226/24862.

The National Academies of
SCIENCES · ENGINEERING · MEDICINE

The **National Academy of Sciences** was established in 1863 by an Act of Congress, signed by President Lincoln, as a private, nongovernmental institution to advise the nation on issues related to science and technology. Members are elected by their peers for outstanding contributions to research. Dr. Marcia McNutt is president.

The **National Academy of Engineering** was established in 1964 under the charter of the National Academy of Sciences to bring the practices of engineering to advising the nation. Members are elected by their peers for extraordinary contributions to engineering. Dr. C. D. Mote, Jr., is president.

The **National Academy of Medicine** (formerly the Institute of Medicine) was established in 1970 under the charter of the National Academy of Sciences to advise the nation on medical and health issues. Members are elected by their peers for distinguished contributions to medicine and health. Dr. Victor J. Dzau is president.

The three Academies work together as the **National Academies of Sciences, Engineering, and Medicine** to provide independent, objective analysis and advice to the nation and conduct other activities to solve complex problems and inform public policy decisions. The National Academies also encourage education and research, recognize outstanding contributions to knowledge, and increase public understanding in matters of science, engineering, and medicine.

Learn more about the National Academies of Sciences, Engineering, and Medicine at **www.nationalacademies.org**.

The National Academies of
SCIENCES • ENGINEERING • MEDICINE

Consensus Study Reports published by the National Academies of Sciences, Engineering, and Medicine document the evidence-based consensus on the study's statement of task by an authoring committee of experts. Reports typically include findings, conclusions, and recommendations based on information gathered by the committee and the committee's deliberations. Each report has been subjected to a rigorous and independent peer-review process and it represents the position of the National Academies on the statement of task.

Proceedings published by the National Academies of Sciences, Engineering, and Medicine chronicle the presentations and discussions at a workshop, symposium, or other event convened by the National Academies. The statements and opinions contained in proceedings are those of the participants and are not endorsed by other participants, the planning committee, or the National Academies.

For information about other products and activities of the National Academies, please visit www.nationalacademies.org/about/whatwedo.

**COMMITTEE ON REDUCING THE THREAT OF
IMPROVISED EXPLOSIVE DEVICE ATTACKS BY
RESTRICTING ACCESS TO CHEMICAL EXPLOSIVE PRECURSORS**

Members

VICTORIA A. GREENFIELD (*Chair*), George Mason University
ROBERT G. BEST, Defense Threat Reduction Agency – JIDO
LEO E. BRADLEY, LE Bradley Consulting LLC
JOHN C. BRULIA, Austin Powder Company (Retired)
CARRIE L. CASTILLE, Independent Consultant
DAVID G. DELANEY, University of Maryland
ARTHUR G. FRAAS, Resources for the Future
WILLIAM J. HURLEY, Institute for Defense Analysis
KARMEN N. LAPPO, Sandia National Laboratories
BECKY D. OLINGER, Los Alamos National Laboratory
JIMMIE C. OXLEY, University of Rhode Island
KEVIN F. SMITH, Sustainable Supply Chain Consulting
KIRK YEAGER, Federal Bureau of Investigation

Staff

CAMLY TRAN, Study Director
SAMUEL M. GOODMAN, Postdoctoral Fellow
JARRETT I. NGUYEN, Senior Program Assistant

BOARD ON CHEMICAL SCIENCES AND TECHNOLOGY

Co-Chairs

DAVID BEM, PPG Industries
DAVID R. WALT, Tufts University

Members

HÉCTOR D. ABRUÑA, Cornell University
JOEL C. BARRISH, Achillion Pharmaceuticals, Inc.
MARK A. BARTEAU, NAE, University of Michigan
JOAN BRENNECKE, NAE, University of Notre Dame
MICHELLE V. BUCHANAN, Oak Ridge National Laboratory
DAVID W. CHRISTIANSON, University of Pennsylvania
JENNIFER SINCLAIR CURTIS, University of California, Davis
RICHARD EISENBERG, NAS, University of Rochester
SAMUEL H. GELLMAN, NAS, University of Wisconsin–Madison
SHARON C. GLOTZER, NAS, University of Michigan
MIRIAM E. JOHN, Sandia National Laboratories *(retired)*
FRANCES S. LIGLER, NAE, University of North Carolina at Chapel Hill and North Carolina State University
SANDER G. MILLS, Merck Research Laboratories *(retired)*
JOSEPH B. POWELL, Shell
PETER J. ROSSKY, NAS, Rice University
TIMOTHY SWAGER, NAS, Massachusetts Institute of Technology

National Academies of Sciences, Engineering, and Medicine Staff

TERESA FRYBERGER, Board Director
MARILEE SHELTON-DAVENPORT, Senior Program Officer
CAMLY TRAN, Program Officer
ANNA SBEREGAEVA, Associate Program Officer
SAMUEL M. GOODMAN, Postdoctoral Fellow
JARRETT I. NGUYEN, Senior Program Assistant
SHUBHA BANSKOTA, Financial Associate

Preface

"Jumping to conclusions is efficient if the conclusions are likely to be correct and the costs of an occasional mistake acceptable. Jumping to conclusions is risky when the situation is unfamiliar, the stakes are high and there is no time to collect more information."

Daniel Kahneman, *Thinking, Fast and Slow*

Daniel Kahneman, in *Thinking, Fast and Slow*, distinguishes between deliberative thinking and intuitive thinking, leaving us to consider the benefits of the former and the pitfalls of the latter in policy making. Deliberative thinking can occur proactively, enabling policy makers to weigh tradeoffs, recognize complexity, and focus on long-term strategies for coping with crises; whereas intuitive thinking, which might occur reactively in the aftermath of a crisis, leans toward rapid and simple decisions based on emotion and familiarity.

Although it has been more than two decades since the United States experienced the truck bombings of the Alfred P. Murrah Federal Building in Oklahoma City and the World Trade Center in New York City, terrorist attacks with smaller-scale improvised explosive devices (IEDs) in Paris, France (2015); Brussels, Belgium (2016); New York and New Jersey (2016); and Manchester, United Kingdom (2017) serve as concrete reminders that IEDs remain a persistent threat to the United States and its allies. One could hardly describe this chain of events as a lull in terrorist activity, but the absence of a recent domestic episode like that in Oklahoma City suggests an opening for U.S. policy makers to deliberatively work through some of the most challenging issues around the threat of IEDs, in a period of relative calm.

To that end, the U.S. Department of Homeland Security (DHS) asked the National Academies of Sciences, Engineering, and Medicine to consider opportunities to reduce the threat of IED attacks by restricting access to precursor chemicals that can be used to make homemade explosives. In response, the National Academies assembled a 13-member committee of experts on chemistry, energetic materials, supply chain management, economics, defense, law, and other fields to prioritize the precursor chemicals that can be used to make homemade explosives, to analyze the movement of those chemicals through the domestic supply chain and identify potential vulnerabilities, to examine current domestic and international regulation of the chemicals, and to compare economic, security, and other tradeoffs among potential control strategies. The National Academies selected committee members with backgrounds in research, industry, and policy making and with experience on the ground to ensure that we on the committee considered the scientific, practical, and policy aspects of our findings, conclusions, and recommendations.

We spent much of our time establishing priorities, examining the supply chains of those chemicals that we deemed most concerning, and looking for vulnerabilities as the chemicals make their way to end users. In mapping various regulations and voluntary programs to the chemicals' supply chains to look for gaps, we found more evidence of visibility and oversight, albeit piecemeal, in import, manufacturing, storage, and distribution than in retailing. For that reason, we chose to focus our deliberations on control strategies that could address retail-level vulnerabilities, including those pertaining to e-commerce.

From the outset, we recognized that as long as explosive materials such as black and smokeless powders are readily available, the threat of IED attacks cannot be eliminated; nevertheless, we identified a set of possible control strategies, featuring different types of restrictions on access to precursor chemicals that could play a part in risk reduction. This report considers the benefits, costs, and uncertainties of each approach, but does not provide the comprehensive analysis of specific proposals that would be necessary for policy making.

Drawing inspiration from Kahneman, we argue for treating this report as the starting point of an ongoing deliberative process, which would include a fuller, quantitative analysis of benefits, costs, and uncertainties, not as an end point for decision-making. Even if event-driven policy making is unavoidable, we have tried to lay the groundwork for better policy responses so that in-the-moment decision-making can look more like thoughtful decision-making.

Lastly, I would like to take this opportunity to thank each committee member and the National Academies staff for their contributions and support. It has been an honor to work with such an outstanding group of dedicated individuals.

Victoria A. Greenfield, *Chair*

Acknowledgments

The completion of this study would not have been successful without the assistance of many individuals and organizations. The committee would especially like to thank the following individuals and organizations for their contribution during this study:

U.S. Department of Homeland Security, which sponsored the study and provided valuable information on the agency's responsibilities with the Chemical Facilities Anti-Terrorism Standards (CFATS) program and risk assessment structure. The committee would especially like to thank the director of the Infrastructure and Security Compliance Division, David Wulf, as well as Craig Conklin (Office of Infrastructure Protection) who served as the DHS liaison to the committee and was effective in responding to the committee's requests for information.

Eva-Maria Engdahl, **Ivette Tarrida-Soler**, and **Michael Berglund** of the European Commission and **Anne-Marie Fry** and **Nathan Munson** at the United Kingdom's Home Office for hosting two members of the committee and a staff officer to discuss their current regulations on chemical explosive precursors.

Speakers and invited participants at the committee's data-gathering meetings. These individuals are listed here: Andy O'Hare, The Fertilizer Institute; Cynthia Hilton, The Institute of Makers of Explosives; Tony Cheesebrough, Tom Colley, Todd Klessman, Kelly Murray, Mike Pickford, and Patrick Starke, U.S. Department of Homeland Security; Matt Hendley and Kevin Sheehan, U.S. Department of Justice – FBI; Col. Bradley B. Preston, U.S. Department of Defense – Joint Improvised-Threat Defeat Organization; Special Agent Will McCray, U.S. Department of Justice – Bureau of Alcohol, Tobacco, Firearms, and Explosives;

William Hoffman, U.S. Department of Agriculture; James Bevan, Conflict Armament Research; Noel Hsu, Orica; Donald Thomas, CF Industries; Gary Vogen, Yara; Hank Sattlethight, The Aluminum Association; Ross Anderson, Arkema; Chris Gibson, Hawkins Inc.; Julie Heckman, American Pyrotechnic Association; Jennifer Gibson, National Association of Chemical Distributors; David Closs, Michigan State University; Henry Willis, RAND Corporation; Lisa Robinson, Harvard University; Michael Lewis, ANGUS Chemical Company; Kris Griffith, American Pacific Corporation; Daniel Roczniak, American Chemistry Council; Steven Krupinsky, U.S. Customs and Border Protection; Antonio Guzman, U.S. Drug Enforcement Administration; LCDR Adam Cooley and Betty McMenemy, U.S. Coast Guard; Lisa Long and Jeffrey Wanko, U.S. Occupational Safety and Health Administration; Paul Bomgardner and Steven Webb, U.S. Department of Transportation; Philip Davison, Association of American Plant Food Control Officials; Thomas Farmer, Association of American Railroads; Boyd Stephenson, National Tank Truck Carriers; Kyle Liske, Agriculture Retailers Association; Nicholas Cindrich, CVS/Caremark; Howard Kunreuther, University of Pennsylvania; Clare Narrod, University of Maryland; Drew Sindlinger, Nathan Tsoi, and Ramana Kasibhotla, Transportation Security Administration; and Christopher Logue, New York Department of Agriculture.

Acknowledgment of Reviewers

This Consensus Study Report was reviewed in draft form by individuals chosen for their diverse perspectives and technical expertise. The purpose of this independent review is to provide candid and critical comments that will assist the National Academies of Sciences, Engineering, and Medicine in making each published report as sound as possible and to ensure that it meets the institutional standards for quality, objectivity, evidence, and responsiveness to the study charge. The review comments and draft manuscript remain confidential to protect the integrity of the deliberative process.

We thank the following individuals for their review of this report:

RICK BLASGEN, Council of Supply Chain Management Professionals
RUTH DOHERTY, University of Maryland, College Park
JULIE HECKMAN, American Pyrotechnics Association
NOEL IISU, Orica Limited
MICHAEL KENNEDY, Kennedy Law and Policy
RUSSELL MCINTYRE, Defense Intelligence Agency (retired)
ABDUL-AKEEM A. SADIQ, Indiana University–Purdue University
DWIGHT C. STREIT, University of California, Los Angeles
TIM SWAGER, Massachusetts Institute of Technology

Although the reviewers listed above have provided many constructive comments and suggestions, they were not asked to endorse the conclusions or recommendations of this report nor did they see the final draft before its release. The review of this report was overseen by **JOHN ANDERSON**, Illinois Institute of Technology, and **FRAN LIGLER**, North Carolina State University and University of North Carolina Chapel Hill. They were responsible for making certain that an independent examination of this report was carried out in accordance with the standards of the National Academies and that all review comments were carefully considered. Responsibility for the final content of this report rests entirely with the authoring committee and the National Academies.

Contents

SUMMARY   1

1 INTRODUCTION   9
   Charge to the Committee and Interpretation of Scope, 11
   Study Origins, 12
   Analytical Approach, 17
      Framework and Tools, 19
      Data Gathering, 20
      Definitions, 20
   Organization of the Report, 21

2 PRECURSOR CHEMICALS USED TO MAKE HOMEMADE EXPLOSIVES   23
   Past and Recent Attacks Involving Explosives, 23
      Case Study: The Evolving Tactics of a Terrorist Group, 27
   Identifying and Prioritizing Precursor Chemicals Used in IED Attacks, 29
      Charge Size Analysis, 29
      Generating a Short List of Precursor Chemicals, 32
      Criteria for Generating Groups A, B, and C, 32
      Application of the Criteria to Precursor Chemicals, 33
   Conclusion, 34

3 DOMESTIC CHEMICAL SUPPLY CHAIN 37
Supply Chain Overview, 39
  Production and Input Nodes, 42
  Transportation Modes, 43
  Distribution and Retail Nodes, 44
  End User Nodes, 45
  Internet Commerce, 46
Domestic Policy Mechanisms, 48
  Department of Justice, 50
  Department of Homeland Security, 52
  Environmental Protection Agency, 54
  Department of Labor, 54
  Department of Transportation, 55
  Department of Commerce, 56
  State and Local Regulations, 57
  Private-Public Partnerships, 58
  Trade Associations Programs, 59
  Outreach, 61
  Best Practices, 62
Supply Chain Vulnerabilities, 62
  Types of Vulnerabilities, 62
  Coverage of Controls and Other Policy Mechanisms, 64
Exploding Targets, 65
  Chemical Characteristics, 65
  Legal Considerations, 66
Conclusion, 69

4 INTERNATIONAL REGULATIONS 73
Australia, 76
Canada, 77
Singapore, 78
European Union, 78
  The Standing Committee on Precursors, 79
  Compliance Effectiveness, 79
  Regulation Effectiveness, 82
  Challenges and Initial Responses, 83
Global Shield, 83
Conclusion, 86
  Market Level, 86
  Responsible Entities, 86
  Harmonization, 87

5 ASSESSING POSSIBLE CONTROL STRATEGIES   89
Possible Control Strategies, 91
   Retail-Level Controls, 91
   Other Retail-Level Measures and Activities, 92
   Building a Control Strategy from Controls, Measures, and Activities, 93
Assessing Tradeoffs Among Control Strategies, 96
   Assessment of Benefits, 97
   Assessment of Costs, 101
   Consideration of Uncertainties, 104
   Assessments of Other Measures and Activities, 107
   Summary of Assessments and Tradeoffs, 107
Conclusion, 110

6 POTENTIAL APPROACHES TO RESTRICTING MALICIOUS ACTORS' ACCESS TO PRECURSOR CHEMICALS: CONCLUSIONS AND RECOMMENDATIONS   113
Beyond Precursor Chemicals, 114
Recommendations, 114
   Priority Precursor Chemicals, 115
   Strategies at the Retail Level, 116
   Analysis of Control Strategies, 120
   Voluntary Measures, Activities, and Programs, 121
Priority Research Areas, 123
   Data Collection from Incidents Involving Explosives, 123
   Substitute Chemicals, 124
   Standardized Thresholds, 124
   Behavioral Responses, 124
Final Thoughts, 125

REFERENCES   127

APPENDIXES
A **Acronyms**   141
B **Risk and Risk Management**   145
C **History of High-Profile Bombing Attacks**   153
D **Group A Chemical Supply Chains**   155
E **International Questions**   173
F **Training Materials**   175
G **Methods and Limitations of Regulatory Assessment**   179
H **Examples of Retail-Level Control Strategies and Other Measures or Activities**   183
I **Committee Member and Staff Biographies**   191

Summary

Improvised explosive devices (IEDs) are a type of unconventional explosive weapon that can be deployed in a variety of ways, and can cause loss of life, injury, and property damage in both military and civilian environments. Terrorists, violent extremists, and criminals (collectively referred to in this report as malicious actors) often choose IEDs because the ingredients, components, and instructions required to make IEDs are highly accessible. In many cases, precursor chemicals are used to make homemade explosives (HMEs), which are often a component of IEDs. HMEs are defined in this study as explosives produced from precursor chemicals either physically blended or combined in a chemical reaction. Many precursor chemicals are frequently used in industrial manufacturing and may be available for commercial or personal use. Guides for making HMEs and instructions for constructing IEDs are widely available and can be easily found on the internet.

Other countries restrict access to precursor chemicals in an effort to reduce the opportunity for HMEs to be used in IEDs. Although IED attacks have been less frequent in the United States than in other countries, IEDs remain a persistent domestic threat. Restricting access to precursor chemicals might contribute to reducing the threat of IED attacks and in turn prevent potentially devastating bombings, save lives, and reduce financial impacts.

In 1998, the National Research Council published a study titled *Containing the Threat from Illegal Bombings: An Integrated National Strategy for Marking, Tagging, Rendering Inert, and Licensing Explosives and Their Precursors,* which was responsive to a congressional mandate to the Department of the Treasury in the Antiterrorism and Effective Death Penalty Act of 1996. That study, conducted in the aftermath of the World Trade Center and Oklahoma City bombings, focused on precursor chemicals used primarily to produce HMEs for the main charges in large-scale IEDs, consistent with those attacks. It developed a short list of precursor chemicals of greatest concern in that context and made recommendations for limiting and controlling their availability for illegal use. The results of the study were incorporated in a Department of Homeland Security (DHS) rulemaking on Chemical Facility Anti-Terrorism Standards in 2007. In large part, high-profile terrorist incidents like the Oklahoma City bombing also motivated previous efforts to regulate precursor chemicals in the United States, leading policy makers to focus on ammonium nitrate, one of the precursor chemicals used in that incident.

Twenty years later, the threat of IEDs has evolved. Much of the transformation can be attributed to an internet-enabled increase in the availability of materials and know-how, and to differences in other aspects of the threat environment related to actors, motives, and methods.

At the request of DHS, the Academies assembled a committee of experts in chemistry, energetic materials, supply chain management, economics, illicit markets, defense, law, and other fields to prioritize precursor chemicals that can be used to make HMEs; analyze the movement of those chemicals through United States commercial supply chains and identify potential vulnerabilities; examine current United States and international regulation of the chemicals; and compare the economic, security, and other tradeoffs among potential control strategies.

STUDY APPROACH AND CONCLUSIONS

During the course of its deliberations, the committee heard from experts and stakeholders from government agencies, industry (including trade associations), and academia. Experts and stakeholders presented or provided data on the movement of specific chemicals through the supply chain and on related policy mechanisms directly to the committee during one of several data-gathering meetings and in follow-up communications. Additionally, the committee chair, a committee member, and a staff officer conducted a site visit in April 2017 to the European Commission in Brussels, Belgium, and to the United Kingdom's Home Office in London, England, to learn more about international policy on precursor chemicals and potential control strategies. The committee also sourced references from public documents, including the scientific literature and government reports, but did not use controlled or classified materials to construct any aspect of this report. The committee's overall approach is described in greater detail in Chapter 1.

In responding to the statement of task, the committee focused solely on precursor chemicals used to make HMEs. The committee recognizes, however, that the majority of bombing incidents in the United States involve certain explosives—smokeless powder, black powder, flash powder, and pyrotechnic fillers—likely due to their ease of legitimate acquisition. Thus, no strategy for restricting access to precursor chemicals can eliminate the threat of IED attacks as long as these other explosive materials remain accessible.

A wide range of precursor chemicals can be used to make HMEs. To assess the **Precursor Chemicals Used to Make Homemade Explosives** (Chapter 2), the committee assembled a chronological list of incidents that involved explosives, beginning nearly 50 years ago with the first major domestic incident that employed HMEs, which occurred in Sterling Hall at the University of Wisconsin. The data show that large-scale devices, such as the vehicle-borne IEDs (VBIEDS) used in the World Trade Center and Oklahoma City bombings, and smaller-scale devices, such as the person-borne IEDs (PBIEDs) employed more recently in Boston, New York, and New Jersey, pose a significant risk, even if all did not employ HMEs. Past research and regulatory efforts have tended to focus on the threat posed by VBIEDs, but attacks that employ PBIEDs can have and have had serious consequences both domestically and internationally (see Chapter 2).

This study, in contrast to past studies, focused on precursor chemicals that can be used as main charges in VBIEDs or PBIEDs. However, the committee did not assess precursor chemicals that are used only in very small amounts, as in the manufacture of aviation IEDs or detonators, or certain ubiquitous chemicals, such as food products, which would be too difficult to control.

The committee developed a prioritized list of precursor chemicals according to three criteria: (1) whether the precursor chemical could be used in both VBIEDs and PBIEDs, (2) whether the precursor chemical had a history of use in IED attacks, and (3) whether the precursor chemical could be used to make HMEs, independent of the presence of another specific chemical. On that basis, the committee established three groups of precursor chemicals—Groups A, B, and C—by order of priority (see Table S-1). With one exception, it placed chemicals that satisfied all three criteria in Group A, indicating the highest current priority. It placed chemicals that satisfied two of the three criteria in Group B; and it placed chemicals that satisfied one of the three criteria in Group C. The precursor chemicals in Groups B and C do not pose the greatest threat at the time of this report, but could gain greater prominence in the future given the potential for changes in the threat environment and use of the precursor chemicals in HMEs. Because the threat of IED attacks continues to evolve, the committee stresses the importance of reevaluating the groupings to address shifting tactics and precursor chemical use.

The committee next examined the **Domestic Chemical Supply Chains** (Chapter 3) of Group A precursor chemicals, the security mechanisms in place along the supply chains, and potential vulnerabilities that terrorists might exploit to access precursor chemicals. Precursor chemicals enter the supply chain as imports or through manufacturing; are transported and stored; arrive at retail locations; and exit via export or sale to consumers or to agricultural or industrial end users. Numerous federal agencies (including DHS, DOJ, DOT, EPA, and OSHA), state agencies, and voluntary programs, like ResponsibleCare and

TABLE S-1 Ranking of Precursor Chemicals into Three Groups

		Charge Size	Prior Use	Dependency
Group A	Aluminum (powder, paste, flake)	V/P	Y	I
	Ammonium nitrate	V/P	Y	I
	Calcium ammonium nitrate	V/P	Y	I
	Hydrogen peroxide	V/P	Y	I
	Nitric acid	V/P	Y	I
	Nitromethane	V/P	Y	I
	Potassium chlorate	V/P	Y	I
	Potassium perchlorate	V/P	Y	I
	Sodium chlorate	V/P	Y	I
	Urea ammonium nitrate solution	V/P	N*	I
Group B	Calcium nitrate	V/P	N	I
	Hydrochloric acid	V/P	N	I
	Potassium nitrate	V/P	N	I
	Potassium permanganate	P	Y	I
	Sodium nitrate	V/P	N	I
	Sodium nitrite	P	Y	I
	Sulfur	V/P	N	I
	Sulfuric acid	V/P	Y	D
	Urea	V/P	Y	D
	Zinc (powder)	P	Y	I
Group C	Ammonium perchlorate	P	N	I
	Antimony trisulfide	P	N	I
	Hexamine	P	Y	D
	Magnalium (powder)	P	N	I
	Magnesium (powder)	P	N	I
	Pentaerythritol	P	Y	D
	Phenol	P	Y	D
	Potassium nitrite	P	N	I

NOTE: *See discussion in Chapter 2 for explanation of including UAN in Group A. V: VBIED (~40 lbs to 10,000 lbs), P: PBIED (~1 lb to 40 lb), Y: used historically, N: not used historically, I: independent, D: dependent.

ResponsibleAg, managed by private organizations contribute to restricting access to precursor chemicals, but they do so differently, resulting in inconsistencies and gaps in oversight. Nevertheless, industry tracks movement of precursor chemicals, especially in bulk quantities, through much of the supply chain, and security mechanisms are in place at many of the early nodes in the chain, starting with imports and manufacturing.

By contrast, the available data suggest that a malicious actor can easily acquire enough precursor chemicals to manufacture a HME through legal purchases at retail outlets, especially outside of agriculture. The precursor chemicals sold at retail outlets have legitimate uses and often fall below the quantity thresholds mandated by existing regulations, and are therefore subject to little or no oversight as a security matter.

The committee concluded that retail-level sales present a substantial vulnerability in the supply chains under consideration. Retailers, especially those selling through internet commerce, have not been a major focus of federal regulation or of voluntary programs, except in a limited number of specific circumstances and sectors, such as agriculture. Internet commerce presents additional challenges because of the anonymity of purchasers, the ease of sharing information, and the large volume of transactions that occur online. All Group A precursor chemicals that can be purchased online can be delivered to a home or business address with very few restrictions.

Presently, binary exploding target kits present one of the most accessible and readily usable forms of precursor chemicals. These kits contain precursor chemicals in the proper weights and physical forms—and with instructions for use—to create an optimized HME that does not require either a commercial detonator or any other primary explosive to detonate. Moreover, exploding target kits are widely available through online and brick-and-mortar retailers.

The committee looked at **International Regulations** (Chapter 4), specifically the regulations in place to restrict access to precursor chemicals in Australia, Canada, Singapore, the United Kingdom, and the European Union (EU) overall, for insight into possible control strategies. The regulators in those locations have tended to focus on point-of-sale restrictions and have accumulated enough experience to provide some information on the strengths and weaknesses of different approaches to restricting access, even if the applicability to the United States' policy environment is imperfect. Although the benefits of security regulations are difficult to track (Appendix G), authorities in the EU believe their restrictions on access, which have included retail bans, licensing, and registries, have reduced the threat of attacks using IEDs made with HME precursor chemicals, albeit at some cost to commerce. They frame threat reduction as the decrease in the amount of explosives precursors on the market and the increase in capacity for law enforcement authorities to investigate suspicious incidents involving explosives precursors.

Using the lessons learned from international regulations, the committee began **Assessing Possible Control Strategies** (Chapter 5) that could include different combinations of mandatory and voluntary policy mechanisms, directed at retail sales to noncommercial end users (i.e., the general public). The committee considered tradeoffs among security, economic, and other factors associated with a small set of possible strategies, drawing notionally from principles of regulatory assessment (Appendix G), including those of benefit-cost analysis. It assessed the strategies qualitatively in relation to three goals: (1) restricting malicious actors' access to precursor chemicals, (2) gathering and disseminating information to prevent or respond to terrorist incidents, and (3) minimizing the burdens on legitimate commerce and use. In deference to the third goal, the committee did not consider restrictions on access for commercial end users; in each case, a commercial purchaser, such as a housecleaning, pool, or spa service or a beautician, would only be expected to provide evidence of commercial status to complete a transaction.

The committee considered four general types of control strategy, each differing by the form of control, defined as a mandatory restriction on access to precursor chemicals. Three of the strategies would feature a new control, either a ban, licensing, or a registry; one of the strategies, referred to here as "business as usual plus," would not feature a new control, but would augment any existing controls with supplemental measures and activities, such as outreach, training, and reporting. Indeed, any of the four strategies, not just business as usual plus, could incorporate such measure and activities. The committee notes that outreach, training, and reporting could be implemented as mandatory or voluntary programs, implying varying degrees of government and industry involvement and the potential for a hybrid strategy.

Overall, of the four types of control strategy considered, none emerged as a best choice during the committee's deliberations on security, economic, and other tradeoffs. For example, the committee found the benefits of stringency might come at the price of forgone sales and use, displacement to other forms of terrorism, and commercial disruption. The committee lacked the time, resources, and directive from the U.S. Department of Homeland Security (DHS) to conduct a comprehensive and detailed analysis of policy options. Thus, this report constitutes a starting point, not an ending point for evaluating possible control strategies.

A full analysis, as would be necessary to support a policy decision, would require more time, data, industry participation, and specificity about the structure and content of proposed actions. Any such analysis should also consider the results of existing domestic programs that restrict access to precursor chemicals, including those intended to curb illicit drug production, and programs adopted in other countries. The EU's continuing experience with bans, licensing, and registration might provide fertile ground for an analysis of those types of controls and lessons learned about what is, or is not, working in the EU, the related costs and unintended consequences, and the applicability to U.S. policy making.

RECOMMENDATIONS

Precursor chemicals have played a role in the manufacture of HMEs in prior IED terrorist attacks, and that role will continue to change with time as the threat environment evolves. Given the ease of accessing HME precursor chemicals and the information to manufacture IEDs, particularly via the internet, the threat of terrorist attacks using IEDs within the United States has not abated.

While the United States has not experienced an attack with a large-scale VBIED since the 1990s, the committee stresses the importance of deliberative thinking before crises, and cautions against intuitive thinking and action bias during and after crises. History suggests that a push for a policy response—from the public and government officials—often follows a crisis, but event-driven policy making, based on spontaneous, intuitive thinking, suffers limitations and can yield counterproductive results. The committee also stresses the importance of periodically reevaluating priorities among precursor chemicals, in light of changes in the threat environment, and of designing control strategies with means of harvesting and leveraging data to learn from experience and strengthen the strategies over time.

Pursuant to the primary goal of reducing the threat of IED attacks by restricting access to precursor chemicals, the committee details six recommended courses of action and four research areas meriting future attention in Chapter 6. The committee emphasizes the essential role of Congress in developing and implementing appropriate risk-reducing control strategies. Congress can be particularly instrumental in ensuring that crisis-driven interests do not unduly influence new laws or regulations, and can play a role in each of the six recommendations listed below to enhance the nation's domestic and international risk-reduction programs.

Priority Precursor Chemicals

Recommendation 1: Federal, state, local, and private sector entities attempting to reduce the threat of IED attacks by restricting access to precursor chemicals should focus on both person-borne and vehicle borne IEDs.

Recommendation 2: Federal, state, local, and private sector entities attempting to reduce the threats from person-borne and vehicle-borne IEDs should consider multi-chemical, rather than single-chemical, strategies.

Strategies at the Retail Level

Recommendation 3: Federal, state, local, and private sector entities attempting to reduce the threats from person-borne and vehicle-borne IEDs should focus on retail-level transactions of precursor chemicals, especially e-commerce.

Recommendation 4: Federal, state, local, and private-sector entities should explore strategies for harmonizing oversight of the sale and use of commercially available kits that contain precursor chemicals that are specifically designed to be combined to produce homemade explosives.

Analysis of Control Strategies

Recommendation 5: U.S. DHS should engage in a more comprehensive, detailed, and rigorous analysis of specific provisions for proposed mandatory and voluntary policy mechanisms to restrict access to precursor chemicals by malicious actors.

Voluntary Measures, Activities, and Programs

Recommendation 6: The federal government should provide additional support for voluntary measures, activities, and programs that can contribute to restricting access by malicious actors to precursor chemicals used to manufacture IEDs.

RESEARCH PRIORITIES

In addition to the aforementioned recommendations, the committee identified several areas of research that could provide additional pathways for limiting access to precursor chemicals or designing appropriate regulations. Major topics for future research include the following:

- standardization of explosives incident data collection;
- substitution of precursor chemicals in commercial products;
- standardization of regulatory thresholds; and
- understanding behavioral responses, including those of terrorists, to proposed controls and those of policy makers to terrorist attacks.

1

Introduction

An improvised explosive device (IED), as defined in this report, is "a device placed or fabricated in an improvised manner incorporating destructive, lethal, noxious, pyrotechnic or incendiary chemicals and designed to destroy, incapacitate, harass or distract. It may incorporate military stores, but is normally devised from non-military components."[1,2] IEDs can vary widely in size, but most have at least four components: a power source, an initiator, a main charge, and a switch (see Figure 1-1). In addition, an IED can employ a casing or container, a booster, and added fragmentation or shrapnel, such as nails or screws. None of the four core components are difficult to obtain, and some, such as batteries, switches, and even certain explosive materials, are ubiquitous. The main charge can include homemade, military, or commercial explosives, or pyrotechnics and propellants (see Figure 1-1).

While most explosives incidents reported in the United States involve commercial explosives (e.g., dynamite) or pyrotechnics (e.g., fireworks, flash powder) and propellants (e.g., black and smokeless powders),[3] there are also consistent reports of homemade explosives (HMEs) derived from precursor chemicals. HMEs are defined in this study (based on United Nations [UN] literature[1]) as explosives produced from precursor chemicals either physically blended or combined in a chemical reaction.

The United States has not experienced an attack with a large-scale IED since the truck bombings of the World Trade Center in New York City in 1993 and the Alfred P. Murrah Federal Building in Oklahoma City in 1995, but more-recent attacks with smaller-scale IEDs, including the bombings in Boston (2013) and New York and New Jersey (2016), starkly demonstrate the ongoing threat of IEDs, even if all did not employ HMEs. The Boston attack, which involved

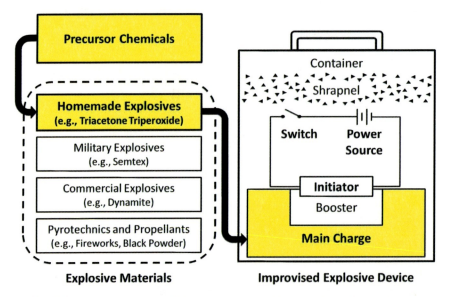

FIGURE 1-1 Components, possible explosive materials, and the position of precursor chemicals in an IED. The IED core components are the power source, initiator, main charge, and switch. Optional components consist of a booster charge, shrapnel, and a container. While drawn as a switch and power source, the initiating mechanism is not limited to electrical systems.

pyrotechnic filler, resulted in three deaths and more than 260 injuries, and millions of dollars in economic losses.[4]

Twenty years after the Oklahoma City bombing, the ingredients, components, and instructions for producing IEDs remain accessible to terrorists, violent extremists, and criminals—collectively referred to in this report as *malicious actors*—who are intent on inflicting casualties,[5] damaging critical infrastructure,[6] and eliciting fear. Instructional videos and how-to guides for making IEDs are readily available and transmitted via the internet and in-person meetings between would-be bombers and their mentors are facilitated using modern telecommunications and global travel.[7] Precursor chemicals can be obtained legally from brick-and-mortar retailers (e.g., hardware, drug, and garden supply stores) or online, because of their widespread legitimate uses. IED attacks in France,[8] Belgium,[9] and the United Kingdom[10] provide further evidence of IEDs' global relevance. Criminal enterprises and homegrown extremists may use IEDs less frequently in the United States than elsewhere in the world, but the consequences can be serious.

CHARGE TO THE COMMITTEE AND INTERPRETATION OF SCOPE

At the request of the U.S. Department of Homeland Security (DHS), the National Academies of Sciences, Engineering, and Medicine (the National Academies) assembled an ad hoc committee to identify and prioritize a list of precursor chemicals that have been used or are susceptible for use in IEDs in the United States or internationally; to analyze the movement of priority precursor chemicals through commercial supply chains and identify potential vulnerabilities; and to suggest control strategies, in light of current U.S. and international policy and the various tradeoffs among mandatory and voluntary approaches (Box 1-1).[11] As a consequence, the focus of this report is not on the threat of IED attacks overall, but on the accessibility of precursor chemicals to malicious actors.

BOX 1-1
Statement of Task

To assist the Department of Homeland Security (DHS) in its efforts to safeguard the Nation's people, infrastructure, and economy from terrorist use of improvised explosive devices (IEDs), the National Academies of Sciences, Engineering, and Medicine (the Academies) will identify priority precursor chemicals used in the manufacture of IEDs, suggest controls that could be considered as part of a voluntary or regulatory scheme, and examine tradeoffs among those strategies. The committee will:

1. Review the available literature and data, both United States and international, to identify and list chemicals that have been used either in the United States or internationally or are susceptible for use in IEDs. For chemicals found to be currently used in IEDs, identify these chemicals in order of the most widely used to the least used.
2. For each of the listed chemicals, analyze how the chemical moves through commercial supply chains. Assess the vulnerabilities and weaknesses of the supply chains with respect to susceptibility to theft and diversion of the chemical.
3. Using the information obtained in Steps 1 and 2 to develop a priority ranking of chemicals to consider for control and discuss the criteria used.
4. Describe and assess existing control measures, both in the United States and internationally, for the priority chemicals, including vulnerabilities in the existing framework of voluntary and regulatory controls.
5. Suggest controls that might be effective for a voluntary or regulatory strategy and discuss the tradeoffs between factors such as economics, cost, security, and impact on commerce.

The report will provide a prioritized list of precursor chemicals that appear to pose the greatest risks in terms of their potential for use in the manufacture of IEDs and strategies for voluntary and regulatory control.

The committee was directed to focus its assessment solely on precursor chemicals used to make HMEs, but recognizes that a significant number of incidents in the United States used commercial explosives and smokeless, black, and flash powders and pyrotechnic fillers applied as the explosive materials in IEDs (see Chapter 2).[3] Thus, no strategy for restricting access to precursor chemicals can eliminate the threat of IED attacks as long as these other explosive materials remain accessible.

The committee prioritized precursors from an expansive list of chemicals (see Chapter 2 for more details), examined all levels of the chemicals' supply chains for potential vulnerabilities, explored domestic and international policy for insight, and considered security, economic, and other tradeoffs among strategies featuring different combinations of mandatory and voluntary policy mechanisms. Cognizant of the importance of preserving legitimate commerce and use, the committee explored the possibility of strategies that could include new mandatory restrictions on access to precursor chemicals, but it did not presume the necessity of new mandatory restrictions. It also considered other measures and activities, including outreach, training, and reporting, that could supplement new restrictions or be applied independently. Such measures and activities could operate through government channels, industry-led initiatives, or both, under legal mandates or through voluntary participation.

Box 1-2 describes the use of IEDs by terrorist networks. The statement of task pertains to HMEs, specifically the precursor chemicals used to make them, and to concerns that are embedded in the highlighted boxes—labeled Gather and Provide Resources and Improvise Concepts of Operations, Tactics, and Devices—in the threat chain depicted in Figure 1-2. On the basis of the committee's limited focus on precursor chemicals, and given the prevalence of alternative explosive materials (such as propellants and pyrotechnics) in domestic incidents involving explosives (see Chapter 2), the strategies under consideration in this report might play a part in reducing the likelihood of malicious actors gaining access to precursor chemicals, but cannot be expected to prevent all malicious actors from manufacturing IEDs or taking other harmful actions.

STUDY ORIGINS

Shortly after the World Trade Center and Oklahoma City bombings, Congress mandated a broad study of issues related to detecting, rendering inert, and licensing precursor chemicals and explosives, which was undertaken by the Academies' Committee on Marking, Rendering Inert, and Licensing of Explosive Materials.[13,14] That effort, which focused on precursor chemicals used primarily to produce HMEs for the main charges in large-scale IEDs, consistent with those attacks, yielded the following results that are of particular relevance to the present study:

- a prioritized list of precursors based on availability and criticality in building an effective IED, with the highest priority chemicals including ammonium nitrate, nitromethane, sodium nitrate, potassium nitrate, sodium chlorate, potassium chlorate, potassium perchlorate, nitric acid, and hydrogen peroxide;
- a recommendation to include procedures to easily amend the precursor list based on ongoing threat assessments given that bombing tactics will change over time;
- recommendations for controlling the prioritized materials based on the perceived threat level—ranging from status quo to greatly increased—through increased awareness, record keeping, licensing, adulteration, or banning; and
- a recommendation for further analysis of the costs of possible actions prior to implementation of any controls and with full participation of affected industries.

In the intervening decades, concerns about the use of IEDs in terrorist attacks have contributed to or driven regulatory action in the United States and abroad (see Figure 1-3).[15-17] The United States (see Chapter 3) and other countries (see Chapter 4) have taken different approaches, with the subjects of concern ranging from high-volume commerce to small, individual transactions.

In 2007, Congress mandated that DHS implement the Chemical Facility Anti-Terrorism Standards (CFATS) program, which identifies and regulates high-risk chemical facilities to ensure they have security measures in place to guard against theft and public health hazards.[17,18] CFATS focuses entirely on facilities and storage and does not cover chemicals in transport. Some of the precursor chemicals that are controlled under CFATS, and the justification for their presence, resulted from the inclusion of the chemicals on the short list that was recommended by the 1998 committee.[14]

In 2008, Congress mandated the expansion of security controls to ammonium nitrate (AN), and DHS proposed supporting regulatory action.[19-21] AN, which has a legitimate role in agricultural, industrial, and other applications, has gained notoriety in the United States and internationally for its use in bombings with large-scale IEDs, including those employed in the bombing of Sterling Hall at the University of Wisconsin[22] in 1970, which was the first major domestic incident to employ HMEs, and later in Oklahoma City.[23] In 2011, a Norwegian extremist acquired AN under the pretense of agricultural use and employed it in an explosion that killed eight and injured more than 200 in Oslo.[24] Some countries, including Northern Ireland[25] and Turkey,[26] have banned AN from their markets. The proposed U.S. Ammonium Nitrate Security Program (Box 1-3) would introduce a registration regime throughout the United States' supply chain and set requirements for transactions, but efforts to complete the rulemaking process have been under contention for almost a decade.

> **BOX 1-2**
> **IEDs as Network Threats**
>
> IEDs have been the leading cause of casualties for U.S. and coalition military forces fighting in the Middle East, and are used against the local population as a way to instill fear and to maintain control.[12] Their global proliferation and employment demonstrates a commitment by transnational terrorist networks to such strategies, and it is imaginable that those networks and others will bring IED-based tactics to the United States. This includes both individual bombers who use available information to act independently, and organized groups or networks of terrorists.
>
> Any group that systematically employs IEDs must perform a series of networked operations or activities to be successful. These operational requirements include (1) funding and (2) resourcing both materials, including explosive materials, and personnel, such as bomb makers who are driven or guided by ideological, economic, or other motives. Network activities may occur simultaneously or sequentially (Figure 1-2), but oftentimes, whether simultaneous or sequential, they occur independently. Each of the functions may be organized as one or more cells within the network, and participants in each function or cell may be unaware of the others' existence.
>
> Each step along the threat chain presents an opportunity to disrupt the network and impair its operations, directly or indirectly. A direct intervention might block access to funding, materials, or personnel, including bomb-making materials and bomb makers. Less directly, information derived from critical and noncritical functions can lead to the identification, tracking, and exploitation of the network's vulnerabilities, contributing to both its disruption and to the impairment of attacks. Through detecting, tracking, and exploiting both general IED and other illicit activities, including the diversion, theft, or other misappropriation of precursor chemicals, government agencies can determine how a network is structured and sustained, and how it conducts its operations.
>
> But, attempts to disrupt and impair can also lead to displacement. If, for example, an organization cannot access a chemical of preference, it might switch from that

It is difficult to make concrete conclusions on the global use of HMEs because the available data are sparse and lexicons are inconsistent. The 2015 U.S. Bomb Data Center (USBDC) report on explosives incidents indicates use of—or intent to use—military explosives, commercial blasting agents, commercial pyrotechnics and propellants, and a variety of HMEs (Box 1-4).[3] AN is just one of several precursor chemicals that have been used in IEDs.

In 2014, European Union (EU) Regulation No 98/2013 entered into force throughout the EU with a goal of enhancing protections to citizens from the

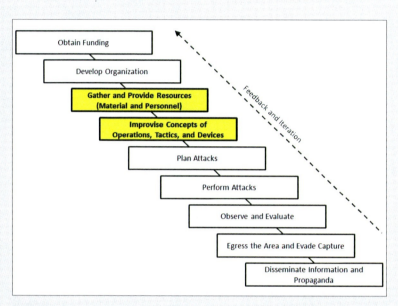

FIGURE 1-2 The IED threat chain.[2] Highlighted boxes indicate portions of the threat chain addressed in this report.

chemical to a different chemical; from HMEs to other common forms of explosives; or from IEDs to other means of attack.

While beyond the scope of this report, identifying and understanding the network's structure and behavior, particularly its potential responses to policy interventions, is a primary step to understanding and countering the IED threat (see Appendix G).

threat of terrorism.[28] It introduced a common framework for controlling access to, introducing, and possessing and using certain substances or mixtures that could be used to make HMEs. Members of the general public have restricted access to and use of the chemicals listed in Annex I of the regulation. The regulatory default is a ban on sales to the general public, but any of the 28 EU member states (MS) may grant access through licenses or registration. In addition, the regulation also introduces rules for retailers who market such substances (see Chapter 4).

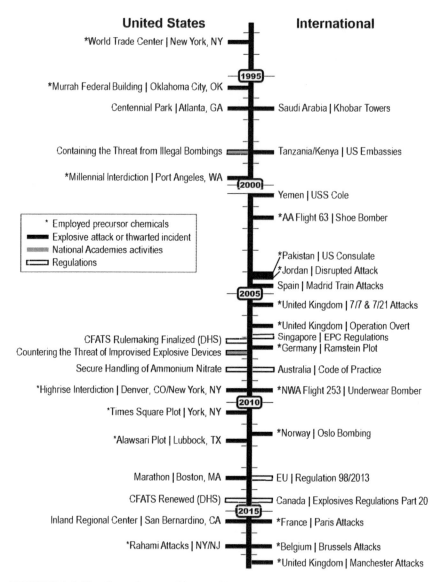

FIGURE 1-3 Timeline of recent, high-profile bombings in the United States and internationally. Explosive events are shown in black, related National Academies activities in grey, and regulatory events in white. Events that involved precursor chemicals are starred.

> **BOX 1-3**
> **Ammonium Nitrate Security Program (ANSP)**
>
> **Motivator:** After the deadly Oklahoma City bombing in 1995, congressional attention was placed on the vulnerability of AN. AN is a chemical that exists in multiple concentrations and physical forms, and principally is used as an agricultural fertilizer, as a component in the manufacture of some first-aid products (such as cold packs), and as a component of explosives often used in the mining and construction industries. In addition to its many legitimate uses, AN was a primary chemical used in the 1995 Murrah Federal Building bombing.
>
> **Tasking:** In 2008, Subtitle J, Section 563 of the Fiscal Year 2008 Department of Homeland Security Appropriations Act amended the Homeland Security Act of 2002 to "regulate the sale and transfer of ammonium nitrate by an ammonium nitrate facility . . . to prevent the misappropriation or use of ammonium nitrate in an act of terrorism." Pursuant to Subtitle J, DHS was instructed to develop a plan to implement Subtitle J that required, at a minimum, the following activities: registration applications by certain AN sellers and prospective AN purchasers, Terrorist Screening Database checks of prospective applicants, registration numbers issued or denied, purchaser verification activities at the point of sale, record keeping at all AN facilities, reporting theft or loss of AN, inspections and audits by DHS for regulatory compliance, guidance materials and posters with relevant information, establishing the threshold percentage of AN in a mixture to be regulated, an appeals process for denied applicants, and penalties for violations.
>
> **Results:** On August 3, 2011, DHS published the ANSP Notice of Proposed Rulemaking.[19]
>
> As part of this process, DHS organized public meetings to solicit information in four key areas: AN use and characteristics; registration; verification; and record keeping.[27] DHS received many comments from the public, including industrial entities. ANSP has not progressed beyond this stage at the time of this study.

The charge to the committee appears to flow naturally from past events and policy developments, both in the United States and internationally. Although the statement of task focuses on a single dimension of the threat of IED attacks, namely access to precursor chemicals, it does so comprehensively, seemingly reflective of past and ongoing efforts to increase security.

ANALYTICAL APPROACH

History, including that of IED attacks, suggests that a push for a policy response—from the public and from government officials—often follows a crisis, but event-driven policy making based on spontaneous *intuitive thinking* might

> **BOX 1-4**
> **Limitations of Available Bombing Data**
>
> A critical component of evaluating the current threat of IEDs containing HMEs made from precursor chemicals is knowing the trends in materials and devices used by bombers. Detailed information in this area is not currently accessible by the committee due to its inherently sensitive nature. The only publicly available data sets are reported in the annual USBDC reports, which provide high-level information on all types of bombings. The utility of the data is limited for the purposes of this report because of several factors:
>
> - the data are voluntarily reported to the Bureau of Alcohol, Tobacco, Firearms and Explosives (ATF) from state and local law enforcement and, therefore, may not provide a complete representation of domestic bombings;
> - the data lack the detailed forensic information needed to compare and collate the precursor chemicals used in incidents involving HMEs, especially non-recovery entries;
> - statistics derived from the data set will be diluted by entries involving materials and devices not under consideration in this report, for example, over-pressure devices; and
> - the data include a large number of unidentified incidents that do not report the identity of the main charge.
>
> Despite these limitations, the available data do show that commercial explosives, black powder, smokeless powder, and pyrotechnic filler were involved in a significant number of incidents, and their use seems to exceed that of precursor chemical–derived HMEs.

yield less desirable outcomes than policy making based on slower-paced *deliberative thinking*.[29,30] The former tends to be myopic, to operate rapidly, automatically, and effortlessly, and to emphasize simple associations, including emotional reactions, recent past experience, and simple decision rules. The latter allocates attention to effortful and intentional mental activities, in which individuals weigh tradeoffs and recognize relevance, interconnectedness, and the need for coordination, and focuses on long-term strategies for coping with extreme events. Adding to the challenges of making policy in a crisis-ridden environment is the potential for an overreaction to *fearsome risk* or *action bias*: when the prospect of a loss triggers strong emotions, people tend to neglect the probability of an event and exaggerate the benefits of preventative, risk-reducing, or ameliorative measures.[31]

This report provides policy makers with a framework, tools, and information to encourage deliberative thinking, such as that undertaken by this committee, before crises occur and to discourage intuitive thinking and action bias during

INTRODUCTION 19

and after crises. Even if event-driven policy making is unavoidable, it might be possible to lay a foundation for better policy responses, so that in-the-moment decision-making can look more like thoughtful decision-making.

Framework and Tools

In this report, the committee attempts to identify (1) conditions under which malicious actors can legally or illegally obtain particular precursor chemicals to produce HMEs that are used to construct IEDs, both large and small, and (2) the means of reducing the likelihood that these precursor chemicals could fall into the wrong hands, for use in terrorist attacks. The committee's framework for considering supply chains, policy, potential vulnerabilities, and control strategies follows a logical progression, consistent with the statement of task. It starts, in Chapter 2, with a winnowing process by which the committee identified a short list of precursor chemicals of particular concern. After grouping the list of chemicals by priority, it then, in Chapters 3–5, does the following:

- constructs supply chains for the highest priority precursor chemicals;
- characterizes the policy mechanisms pertaining to those chemicals and analogous policies;
- maps out the policy mechanisms in relation to the supply chains to identify potential vulnerabilities;
- examines the design and implementation of international policies on precursor chemicals for insight to possible control strategies; and
- identifies and assesses the tradeoffs among potential control strategies.

In the course of its deliberations, the committee drew implicitly and explicitly from tools of risk management and regulatory assessment (see Appendices B and G).

The committee conceptualized risk in terms of a threatening or hazardous condition, the resulting consequences that can arise from the threat or hazard, and the probability and severity of those consequences. Thus, one might observe that DHS's statement of task speaks largely to probability through the initial likelihood that malicious actors will be able to obtain precursor chemicals. The national security literature provides a simple tool—a matrix that juxtaposes probability and severity—for assessing and prioritizing risk as part of a five-step risk management process (see Appendix B, Figures B-1 and B-3).[32,33] The process calls on the policy community to identify hazards; assess probability and severity; develop controls or other policy mechanisms to mitigate risks; implement policy; and supervise, review, and evaluate, continuously.

To develop controls or other policy mechanisms, the same literature indicates that policy makers should balance risk against costs. Taking a step in that direction, the committee considered tradeoffs among security, economic, and other factors associated with a small set of potential strategies for addressing access to precursor chemicals (see Chapter 5), drawing notionally from principles of regulatory assessment, including those of benefit-cost analysis. Given the limitations on the time, resources, and scope of this project, including access to reliable bombing data, the committee did not undertake a full, formal analysis; rather it presents the results of the qualitative, analytical exercise with which it explored the benefits, costs, and uncertainties of alternative strategies.

Data Gathering

To support its deliberations, the committee gathered information from experts and stakeholders from government agencies, industry (including trade associations), and academia. A list of those experts and stakeholders can be found in the Acknowledgments. Information on the movement of specific chemicals through the supply chain and related policy mechanisms was presented or provided directly to the committee during one of several data-gathering meetings and in follow-up communications. The committee's discussions with agencies, groups, and individuals concerned with precursor chemicals were extremely valuable. Additionally, the committee chair, a committee member, and a staff officer conducted a site visit in April 2017 to the European Commission in Brussels, Belgium, and to the United Kingdom's Home Office in London, England, to learn more about international policy on precursor chemicals and opportunities for risk mitigation. The committee sourced additional references from public documents, including the scientific literature and government reports, but no controlled or classified materials were used to construct any aspect of the report.

Definitions

Many terms used in this report lack universally agreed-upon definitions. For the purposes of this report, the committee sets out its use of some terms that appear across chapters.

The committee defined *improvised explosive device* (IED) as "A device placed or fabricated in an improvised manner incorporating destructive, lethal, noxious, pyrotechnic or incendiary chemicals and designed to destroy, incapacitate, harass or distract. It may incorporate military stores, but is normally devised from non-military components."[1,2]

The committee defined *homemade explosives* (HMEs) as explosives produced from precursor chemicals either physically blended or combined in a chemical reaction.

The committee defined *precursor chemicals* as chemicals that can be used, through blending or chemical reaction, to produce HMEs.

The committee addresses *vulnerabilities* in relation to the potential for unexplained loss, diversion, theft, and other forms of misappropriation along commercial supply chains.

The committee uses the term *displacement* to describe policy-induced shifts in terrorists' strategies and tactics. For example, if a policy interferes with terrorists' access to certain precursor chemicals, they might shift from one precursor chemical to another, from HMEs to other forms of explosive materials, or from IEDs to entirely different methods of attack.

The committee uses the term *mandatory* to describe policy mechanisms that are prescribed in and enforceable under federal, state, or local law or ordinance and the term *voluntary* to describe policy mechanisms that lack the same legal basis and force. Mandates, at whatever level of government, often or even typically involve corresponding regulatory action.

The committee reserves the term *control* for mandatory restrictions on access to precursor chemicals. The terms *measure* and *activity* are applied more broadly, to encompass policy mechanisms, such as outreach, training, or reporting, which could be set up under mandates or through voluntary participation, depending on the specific circumstances under consideration. Similarly, the committee also uses the term *policy intervention* broadly.

The committee uses the term *control strategy* to refer to a package of policy mechanisms that can consist of different combinations of new controls, existing controls, and additional measures and activities, each as defined above. On that basis, a strategy for restricting access to precursor chemicals and reducing the likelihood that terrorists will gain access to them would include some form of control, be it a new control, existing controls, or both, but it could take a hybrid approach, by also including a mix of additional mandatory and voluntary policy mechanisms.

ORGANIZATION OF THE REPORT

Chapter 2 discusses the precursor chemicals that can be used for IEDs, the means by which the committee developed a short list of chemicals of particular concern, and the criteria that the committee used to prioritize the chemicals on the short list. Specifically, it discusses whether a chemical lends itself to producing large-scale vehicle-borne IEDs (VBIEDs), smaller-scale person-borne IEDs (PBIEDs), or both; the history of or potential for use of a precursor; and the independence or dependence of a chemical's role in synthesis, defined as the utility of the chemical in the absence of others. The chapter concludes by categorizing the short list of chemicals into three groups (Groups A, B, and C), with Group A being the highest priority and the primary focus of the in-depth supply chain analysis.

Chapter 3 examines how the precursor chemicals flow from import or manufacturer to end user and characterizes the policy mechanisms that pertain to those precursor chemicals. It uses a generalized supply chain diagram, derived from a series of chemical-specific diagrams (Appendix D), that maps out the common nodes, such as factories, commercial distributors, and retailers. It then presents an overlay of existing policy mechanisms (mandatory and voluntary) and discusses them pursuant to their impact on access to the precursor chemicals. The chapter ends with an assessment of the potential vulnerabilities of the supply chains, defined in terms of susceptibility to unexplained loss, diversion, theft, or other misappropriation, by identifying those nodes with relatively low levels of visibility and oversight.

Chapter 4 reviews international policy on precursor chemicals. Locations of focus include Australia, Canada, Singapore, and the EU (with additional focus on the United Kingdom's implementation strategy). The committee turned to international policy for creative solutions to the potential vulnerabilities that it found in the U.S. supply chain.

Chapter 5 draws from each of the foregoing chapters to introduce a set of possible control strategies, consisting of different combinations of controls and other mandatory and voluntary policy mechanisms, and presents the results of a qualitative assessment of tradeoffs among the strategies. A strategy could include a new control, as defined, or it could supplement existing controls with other measures and activities, subject to varying degrees of government involvement and oversight. Chapter 5 considers the tradeoffs among strategies qualitatively, largely in terms of potential benefits and costs, relative to specific policy objectives, namely restricting access to precursor chemicals, gathering information, and minimizing burdens on legitimate commerce and use.

Chapter 6 presents recommendations and suggestions to stakeholders that oversee or have an interest in the manufacture, transport, sale, or use of precursor chemicals, based on the committee's deliberations and assessment. While these suggestions focus on control strategies to mitigate potential vulnerabilities in the U.S. supply chain, primarily at the retail level, the chapter also highlights other opportunities that came to the fore over the course of the project.

2

Precursor Chemicals Used to Make Homemade Explosives

The number of precursor chemicals that can be used to make homemade explosives (HMEs) is large. To prioritize those chemicals, the committee compiled a long list of precursor chemicals; then it established a short list of chemicals of particular concern; and lastly, it applied a set of criteria to the chemicals on the short list and, according to those criteria, it ranked the chemicals in three separate groups: A, B, and C. The Group A precursor chemicals appear to pose the most immediate threat in terms of their potential for use in improvised explosive devices (IEDs), though shifts in bomb makers' tactics could elevate the status of Group B and C chemicals without warning.

PAST AND RECENT ATTACKS INVOLVING EXPLOSIVES

The committee produced a list of selected explosives incidents, both realized and thwarted, starting with the 1970 Sterling Hall Bombing at the University of Wisconsin (Table 2-1).[34-36] This incident was chosen as a logical starting point as it was the first major attack in the United States that employed precursor chemicals to produce the IED's main charge, specifically, ammonium nitrate (AN) mixed with fuel oil (AN/FO). The main charge of an IED contains the largest amount of explosive; a description of the main charge used in each attack is shown in Table 2-1, along with the estimated mass.

The majority of domestic incidents have used and continue to use commercial explosives, smokeless powder, black powder, flash powder, and pyrotechnic fillers as a main charge likely due to their ease of acquisition (e.g., purchasing 50 pounds of black powder requires no federal license or permit).[37,38] These materials have been used in high-profile incidents like the Boston Marathon bombing.[4] However,

TABLE 2-1 Selected Attacks Involving Explosives from 1970 to 2016

Event (Location)	Main Charge	Mass† (lb)
1970-Sterling Hall Bombing (Madison, WI)	AN/FO	2,000
1983-Beirut Barracks Bombing *Beirut, London)	PETN	20,000
1983-US Embassy Bombings (Beirut, Lebanon)	AN/FO	2,000
1992-St. Mary Axe Bombing (London, United Kingdom)	CAN/IS	2,000
1993-World Trade Center Bombing (New York, NY)	Urea Nitrate	1,200
1993-Bishopsgate Bombing (London, United Kingdom)	CAN/IS	4,000
1995-Oklahoma City Bombing (Oklahoma City, OK)	AN/NM	5,000
1996-Manchester Shopping Mall (Manchester, United Kingdom)	CAN/IS	3,000
1996-South Quay Bombing (London, United Kingdom)	CAN/IS	3,000
1996-Khobar Towers Bombing (Khobar, Saudi Arabia)	C4	20,000
1998-US Embassy Bombings (Tanzania, Kenya)	TNT	2,000
1999-Millennial Bomber Interdiction (Port Angeles, WA)	Urea Nitrate	500
2000-USS Cole Bombing (Aden, Yemen)	Mil. Exp.	1,000
2001-Shoe Bomber (AA Flight 63)	PETN	1
2002-Bali Nightclub Bombing (Bali, Indonesia)	$KClO_3$/S/Al	2,000
2003-Marriott Hotel Jakarta Bombing (Jakarta, Indonesia)	$KClO_3$/S/Al	100
2003-Britsh Consulate Bombing (Istanbul, Turkey)	AN/Al	2,000
2003-Casablanca Bombings (Casablanca, Morocco)	TATP/AN	20
2004-Australian Embassy Attack (Jakarta, Indonesia)	$KClO_3$/S/Al	2,000
2004-US Consulate Failed Attack (Karachi, Pakistan)	CHP/Flour	2,000
2004-Disrupted Jordanian Attack (Amman, Jordan)	CHP/Cumin	10,000
2004-US Embassy Attack (Tashkent, Uzbekistan)	AN/Al	20
2004-Madrid Train Bombings (Madrid, Spain)	Dynamite	20
2005-7/7 Underground Bombing (London, United Kingdom)	CHP/Black Pepper	20
2005-7/21 Bombing (London, United Kingdom)	CHP/Flour	20
2006-Operation Overt (London, United Kingdom)	CHP/Tang	1
2006-Disrupted Plot (Ontario, Canada)	AN/FO	7,000
2007-Disrupted Bomb (Ramstein, Germany)	CHP/Flour	1,000
2008-US Embassy Attack (Sana'a, Yemen)	TNT	100
2009-Underwear Bomber (NWA Flight 253)	PETN	1
2009-Operation Highrise Interdiction (Denver, CO/ New York, NY)	CHP/Flour	10
2010-Printer Bombs (United Kingdom, United Arab Emirates)	PETN	1
2010-Failed Times Square Plot (New York, NY)	AN/IS/Sawdust	100

TABLE 2-1 Continued

Event (Location)	Main Charge	Mass† (lb)
2011-Khalid Ali-M Aldawsari Plot (Lubbock, TX)	Picric Acid	20
2011-Osla Bombing (Oslo, Norway)	AN/FO/CAN/Al/MB	2,000
2002-Aurora Theater Shooting (Aurora, CO)	BP	20
2013-Boston Marathon Bombings (Boston, MA)	Pyrotechnic Filler	20
2015-Paris Attacks (Paris, France)	TATP	20
2016-Brussels Attacks (Brussels, Belguim)	TATP	40
2016-Ahmad Khan Rahami (New York/New Jersey)	AN ET/BP/HMTD	10

NOTE: AN: ammonium nitrate, AN/FO: ammonium nitrate/fuel oil, BP: black powder, CAN: calcium ammonium nitrate, CHP: concentrated hydrogen peroxide, HMTD: hexamethylene triperoxide diamine, IS: icing sugar, NM: nitromethane, PETN: pentaerythritol tetranitrate, TATP: triacetone triperoxide, TNT: trinitrotoluene.
†Upper limit of charge mass.
Gray: event involving precursor chemicals. White: event using commercial or military explosives. Black: event with ambiguous sources.
See Appendix C for an expanded table with boosters and initiators.

as shown in Table 2-1, precursor chemicals have played an important role in many bombing incidents over the past several decades.

Events that occurred prior to the Sterling Hall attack primarily relied on commercial explosives (mainly dynamite), with bombers only adopting precursor-based HMEs once commercial explosives became less accessible.

It would be highly impractical to attempt to compile a list of all explosive attacks over the nearly 50-year span covered by Table 2-1. The committee chose to highlight the events in the table for one or more of three reasons:

- events were either high-profile terrorist attacks that garnered appreciable political or public attention, or struck high-profile U.S. targets outside active war zones;
- events used HMEs; and
- events had reliable forensic data with which to identify the charge.

In the 1970s, a large number of small dynamite bombs (less than 20 lb) were used in the United States. While incidents, such as the Harvey's Casino bombing, that involved dynamite in larger-scale devices garnered significant attention at the time, such incidents are not listed in Table 2-1. Moreover, this list also does not reflect the use of IEDs in active military theaters.

Between the 1970s and 2000, a series of larger vehicle bombs emerged in terrorist attacks with main charges in the thousands of pounds range, but in the following decade, bombs with smaller charges like those seen in the 1970s started to appear again. By the 2010s, the use of HMEs in smaller charges was growing. Similarly, there was a related expansion from fertilizer-based materials to a more diverse range of possible precursor chemicals.

HMEs are produced either by blending or cooking. Blending is the most common form of manufacture, and the simplest, as it requires only physically mixing the precursor chemicals together. To make a blended explosive, at least one precursor chemical must be an oxidizer (a chemical source of oxygen) and one must be a fuel (a chemical or compound that can react with oxygen in a combustion-like process). The blasting agent AN/FO and flash powder are both examples of blended mixtures.

Cooking, a term borrowed from the narcotics enforcement community, is a more complicated manufacturing process to make HMEs wherein multiple precursor chemicals are mixed together and chemically react to form an explosive material. Triacetone triperoxide (TATP), urea nitrate, and ethylene glycol dinitrate (EGDN) are all made through cooking reactions. For many HMEs, more than one synthetic route is possible, involving different precursor chemicals.

Groups involved in explosive attacks and the types of explosives employed by each are shown in Figure 2-1. Both the Unabomber[39] and the Provisional Irish Republican Army (PIRA)[40] represent bombing campaigns with roots traced back to the 1970s, and the Fuerzas Armadas Revolucionarias de Colombia (FARC) has a similarly storied history. The remainder of the groups shown in Figure 2-1 include bomb builders in the Iraq and Afghanistan conflicts as well as the newer factions encountered with the rise of ISIS and other extremists. All of these groups use precursor chemicals to produce their HME charges. History has shown that the tactics developed by groups like Al-Qaeda in the Arabian Peninsula have migrated across the world. For example, the trend of using concentrated hydrogen

FIGURE 2-1 Terrorist groups and their commonly used main charges.

peroxide (CHP) to produce IEDs emerged in Pakistan and rapidly transitioned to Jordan, the United Kingdom, Germany, and, eventually, the United States.[41]

Case Study: The Evolving Tactics of a Terrorist Group

The attempt to solve a problem by making policy in the midst of or in response to a crisis can create even greater difficulties. Perhaps one of the best historical examples of the pitfalls of narrowly focusing on immediate events, at least in the context of precursor chemicals, is that of the response of the United Kingdom to the explosives produced by PIRA during its bombing campaign.[42]

The PIRA bombing campaign began around 1971 and employed devices filled with readily available dynamite stolen from quarries and mines. In parallel, during this time frame in the United States, groups such as Weather Underground, Fuerzas Armadas de Liberación Nacional (FALN), and United Freedom Front (UFF) also conducted many attacks using dynamite. Responding narrowly to these events, both the United Kingdom and the United States increased controls on dynamite. In the United States, bombers migrated to readily accessible low-explosive fillers like black powder and smokeless powder (which remain popular choices to this day). Such materials were not accessible in the United Kingdom, but PIRA was able to obtain farm chemicals to replace the dynamite.

The first chemical PIRA used to produce HME mixtures and replace dynamite was sodium chlorate, a strong oxidizer used as a weed killer. Sodium chlorate was mixed with the energetic fuel nitrobenzene to make small explosive charges. To counter the threat of chlorate explosives, the United Kingdom government mandated the addition of a diluent to weed killer to reduce its explosive potential. After chlorate was no longer an option, PIRA turned to AN. Many farmers in Northern Ireland possessed large quantities of AN as it was a chief fertilizer found in agriculture. In addition, with the heavy equipment required for farming, many of the same farmsteads were equipped with diesel tanks and pumps. This combination made for the logical progression of PIRA developing AN/FO-based IEDs.

The transition to AN/FO-based devices by PIRA from its earlier dynamite and chlorate charges had some logistical and tactical consequences. Unlike dynamite, AN/FO is not cap sensitive (the sensitivity of an explosive to initiation by a #8 detonator), does not function properly in small charges, and requires some confinement to reliably function. As a result, the devices produced from AN/FO tended to be larger than the previous dynamite and chlorate devices, and often incorporated metal containers to produce greater confinement. The net result was larger, fragment-producing bombs. These larger, heavier IEDs had to be delivered by vehicles due to their mass. Thus, efforts to keep terrorists from accessing dynamite and chlorate resulted in PIRA's development of the vehicle bomb.

The United Kingdom, under pressure to address the trend of vehicle-borne IEDs (VBIEDs), passed legislation in 1972 that outlawed the possession of AN fertilizers that contained more than 27.5% nitrogen by mass.[25] To replace the outlawed AN, farmers selected the fertilizer calcium ammonium nitrate (CAN). CAN consisted of AN combined with dolomitic limestone (a blend of calcium and magnesium carbonate). This mixture contained 21% diluent (by weight) to the 79% AN, and was tested and found incapable of being used to produce AN/FO.

It did not take long for PIRA explosives chemists to exploit a simple physical weakness in the new CAN formulation. AN was soluble in water, and the dolomite diluent was not. By mixing the CAN in hot water the AN could be dissolved and separated from the insoluble carbonate component. Once the solid was filtered out, the remaining liquid could be driven off to isolate nearly pure AN. During this time period, United Kingdom authorities came across caches of AN in three purity ranges (100%, 80–90%, and 60%). It is notable that the 60% AN product was actually more dilute than the CAN the terrorists were trying to pull AN out of. The use of CAN in farming did not stop PIRA, but it did make the production of AN-based devices more time consuming and removed the least-adept bomb makers from the picture. Thus, the countermeasure had some limited effect.

Initially, the AN recovered from the recrystallization process was not ideal for AN/FO production. It was coarse and crystalline and would not absorb an optimum amount of diesel. To compensate for this change PIRA began using alternative fuels. One very popular formulation developed was a mixture of AN and nitrobenzene (referred to as ANNIE).

In 1991, approximately 19 years after its introduction, PIRA discovered that crushing the CAN prills into a powdered form using either industrial strength coffee grinders or barley crushers eliminated the need to isolate purified AN. The pulverized CAN could be mixed with a variety of fuels to make an effective explosive filler. Two fuels surfaced as constants: aluminum powder and powdered (icing) sugar. Aluminum was applied consistently for smaller, mortar-borne charges, and sugar was used in the larger-scale VBIEDs. During the 1990s, PIRA perfected the CAN and icing sugar mixture and used it in four major bombings in England. Three of these bombs were deployed against the city of London, and one the city of Manchester. The largest was approximately 4,000 pounds (roughly equivalent to the bomb used in Oklahoma City).

The Taliban's development of explosives in Afghanistan in recent years and PIRA's development of countermeasures to overcome attempts to regulate precursor chemicals in the United Kingdom bear striking similarity. The Taliban conducted the same processing operations to weaponize AN as PIRA. However, the Taliban developed its methods in the span of years instead of the decades it took PIRA.

PIRA's transition from commercial dynamite to a variety of AN-based HME mixtures resembles the paths of other determined terrorist groups. Initially, groups attempt to procure commercial or military explosives if such are accessible. In the absence of available explosives, they look for materials that can be blended together, such as to make AN/FO. Denied the precursors for simple blends, they resort to processing materials to produce the feedstock of their explosives, such as by isolating AN from CAN. With each level of difficulty introduced into the process, fewer bombers will be successful in their endeavors. However, any government creating controls for precursor chemicals must consider the tactics that will be developed in response.

IDENTIFYING AND PRIORITIZING PRECURSOR CHEMICALS USED IN IED ATTACKS

Precursor chemicals used to produce HMEs for IEDs can be categorized by type and role as oxidizers, fuels (organic materials, energetic organic compounds, food products, or inorganic materials), and synthesis chemicals (including strong and weak acids; Figure 2-2). The figure, which constitutes the committee's "long list" of precursor chemicals, is not exhaustive, as it would be impossible to list every precursor chemical that has been or can be used in an IED.

Charge Size Analysis

Not all precursor chemicals can be used to make the main charges for every bombing scenario. Figure 2-3 summarizes the various precursor chemicals seen as the main charges for different use-cases: VBIEDs, person-borne IEDs (PBIEDs), aircraft bombings, and detonators. These are not the only possible charges for each use-case.

VBIEDs use charges ranging in mass from approximately 40 pounds to tens of thousands of pounds, depending on the carrying capacity of the vehicle. Precursor chemicals used to produce these explosives tend to be fertilizers (e.g., AN and urea), potassium chlorate, and CHP, given the ability to amass these precursor chemicals in large quantities.

PBIEDs are typically encountered in backpacks, brief cases, small bags, and suicide bombing vests, belts, etc. The charge mass of these devices is predicated on what the individual delivering the charge is capable of carrying. Historically, the charge mass for PBIEDs ranges from approximately 1 to 40 pounds. PBIEDs typically also employ a mass of fragmentation material, such as nails or screws, that can weigh as much as the explosive charge itself.

Explosives used against aviation targets historically have been military formulations due to their reliability and power, although recent terrorist plots against aircraft have used HMEs, albeit below the mass seen in PBIEDs. Terrorists use precursor chemicals frequently in detonator construction, but they also opt for pre-made systems acquired from commercial sources when possible. Detonators use precursor chemicals in very small amounts, but the primary explosives they produce are often very sensitive and unstable. Thus, there is an inherent danger in making, handling, transporting, and storing improvised detonators.

Synthesis Chemicals	Oxidizers	
Acetone	Hypochlorite Salts (Ca^{2+}/Na^+)	
Aspirin	Chlorate Salts (Na^+/K^+)	
Erythritol	Hydrogen Peroxide, Concentrated (CHP)	
Ethylene Glycol	Metal Peroxides (Ba^{2+}/Na^+)	
Glycerol	Nitrate Salts ($Ca^{2+}/Na^+/K^+/NH_4^+$[AN]/$Ca^{2+}NH_4^+$[CAN])	
Hexamine	Nitrite Salts (Na^+/K^+)	
Hydrazine	Perchloric Acid	
Hydrogen Peroxide, Dilute	Perchlorate Salts ($Na^+/NH_4^+/K^+$)	
Mannitol	Potassium Permanganate	
Methanol	**Fuels**	
MEK	Organic Materials	Food Products
Pentaerythritol	Diesel	Artificial Creamer
Phenol	Kerosene	Black Pepper
Sodium Azide	Mineral Oil	Black Seed
Urea	Motor Oil	Cinnamon
UAN Solution	Sawdust	Cocoa
	Vaseline	Cumin
Strong Acids	Inorganic Materials	Flour
Sulfuric Acid	Aluminum (Al), Powder/Paste	Honey
Hydrochloric Acid	Antimony Trisulfide	Icing Sugar
Nitric Acid	Charcoal	Powdered Drink Mix
Weak Acids	Magnalium Powder	Energetic Organic Compounds
Citric Acid	Magnesium Powder	Nitrobenzene
Acetic Acid	Red Phosphorous	Nitromethane (NM)
Ascorbic Acid	Sulfur	
	Titanium Powder	
	Zinc Powder	

FIGURE 2-2 Long list of precursor chemicals sorted by chemical type and role.
NOTE: Ca^{2+}: calcium; Na^+: sodium; K^+: potassium; Ba^{2+}: barium; NH_4^+: ammonium; AN: ammonium nitrate; CAN: calcium ammonium nitrate.

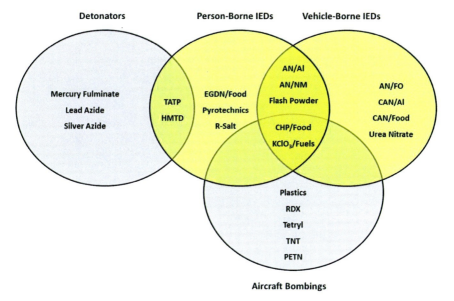

FIGURE 2-3 Historical examples of explosive charges and use-cases.
NOTE: TATP: triacetone triperoxide; HMTD: hexamethylene triperoxide diamine; EGDN: ethylene glycol dinitrate; AN: ammonium nitrate; NM: nitromethane; CHP: concentrated hydrogen peroxide; $KClO_3$: potassium chlorate; R-salt: cyclotrimethylenetrinitrosamine. Food products include flour and icing sugar. For a fuller list of food products, refer to Figure 2-2. Fuels include diesel and saw dust.

Due to the lesser orders of magnitude in aviation IED and detonator charge masses—lesser as compared to the VBIEDs and PBIEDs, described above—the committee limited subsequent analysis to those VBIEDs and PBIEDs, both of which entail sufficient risk to merit consideration.

A scenario involving a larger-scale VBIED, such as a truck bomb, could entail substantially more damage than a scenario involving a smaller-scale PBIED, such as a backpack bomb, but be less likely to occur (Appendix B). Thus, the risk of either scenario might rate concern when both severity and probability are included in the assessment. Starting with these scenarios, one can (1) identify the chemicals that terrorists can use to produce each type of device and the conditions under which they can obtain them; (2) develop strategies to reduce the odds of malicious actors getting access to the precursor chemicals; and (3) ultimately, lessen the risk of either scenario by making both scenarios less likely to happen (i.e., lower probability). While beyond the scope of this study, it may also be possible to drive toward scenarios with less lethal or damaging consequence (i.e., lower severity) by changing access to different precursor chemicals.

Generating a Short List of Precursor Chemicals

Every exercise in prioritization, including this winnowing process, has an inherent degree of subjectivity. Any one of the precursor chemicals listed in Figure 2-2 could be used to produce another devastating attack. To generate a more-focused short list of precursor chemicals, the committee considered two variables: quantity required and ubiquity.

First, it was judged impractical to control very small amounts of any particular precursor chemical and, for this reason, precursor chemicals used only to construct charges for detonators (e.g., mercury and lead azide) and aviation IEDs were eliminated from further consideration.

Second, the committee eliminated certain chemicals on the basis of ubiquity. Ubiquity, for the purposes of this study, described chemicals that are present in high volumes and used in myriad common applications in research, industry, and personal use, such that their analysis by the committee was deemed intractable. All the food products (see Figure 2-2) were removed from consideration because of their ubiquity, as were common hydrocarbons such as diesel fuel. Acetone, however, posed a unique challenge. Acetone is slightly less common than household fuels such as kerosene, but its use in academia and chemical processes makes it one of the most ubiquitous general solvents in the world. While acetone can be reacted with hydrogen peroxide to produce the explosive TATP, the committee did not include acetone on the short list because it is not considered a threat if appropriate steps are taken to control the peroxide component.[14]

By removing precursor chemicals used only in very small amounts and ubiquitous materials, such as food products, the committee narrowed the list of chemicals under consideration to just 28 chemicals, the short list.

Criteria for Generating Groups A, B, and C

To group the short list by priority, the committee adopted three criteria:

- the size of the main charge resulting from the precursor chemical, and whether it can be employed in a VBIED, a PBIED, or both;
- the history of the precursor chemical's use in IED construction; and
- whether the precursor chemical can be used independently, or is dependent on other precursors listed, for the chemical synthesis of an explosive.

Under the first criterion, the committee focused on precursor chemicals that can result in VBIEDs and PBIEDs. These IEDs have charge sizes ranging from several tons to about a pound, as described previously. Some explosives require a large mass to propagate a detonation, and the precursors needed to produce these explosives may not be suited for the production of smaller charges. Other types of explosives are highly susceptible to detonation, making them impractical or difficult to produce at the hundreds of pounds scale; the precursors for these sensitive explosives may be limited to use in smaller quantities. Limitations of precursor availability also dictate usage, independent of the properties of the explosives they can make; some chemicals are simply not available in large quantities. Based on all of these factors, a precursor may have utility in either VBIEDs or PBIEDs, or in both.

Under the second criterion, past usage of a precursor was taken as an indicator of its continued potential to be applied in IEDs in the present and future. Some precursor chemicals have been consistently used in IEDs across the world for many decades, while others have seen only brief use by one isolated terrorist group or individual, only to quickly disappear from malicious use.

Under the third criterion, a precursor chemical merits greater priority if it is independent, that is, if the precursor chemical plays an essential part in the synthesis of an explosive material. For example, as seen in Table 2-1 and Figure 2-1, urea nitrate has been used in HMEs in VBIEDs. To synthesize urea nitrate, the precursor chemicals urea and nitric acid are both required; thus, urea nitrate production could be blocked in the absence of either. Of the two, urea is much more commonly available than nitric acid, and the only explosive it can be used to produce is urea nitrate. In contrast, nitric acid can be used to synthesize a variety of other explosive materials. Thus, in this situation, urea would be categorized as dependent (D) on nitric acid, while nitric acid would be judged independent (I).

Application of the Criteria to Precursor Chemicals

The committee assigned each chemical either a higher or lower priority for each criterion. For the first criterion, chemicals limited to use in either vehicle- or person-borne devices (V or P) were assigned lower priority, while those that could be reasonably anticipated to produce both VBIEDs and PBIEDs (V/P) were assigned higher priority. Aspects discussed earlier, such as the safety and commonality of the chemicals, were considered for this analysis (i.e., whether enough of the final main charge explosive material could be assembled from available materials and without killing the bomb maker).

For the criterion of historical usage, chemicals previously used to produce explosives (Y) were assigned higher priority, and those whose usage was either extremely rare or largely theoretical (N) were assigned lower priority. Ratings for this criterion introduced an element of professional judgment. Every chemical on the list had been used in a bombing or in IED production in some capacity at least once. Ratings were made in a conservative fashion when possible, with some chemicals that had been used by single groups, under very limiting circumstances, receiving a lower priority rating. In some cases, chemicals that had limited past usage were given a higher priority rating due to their versatility and potential for explosives production.

For the third criterion, chemicals judged independent in syntheses (I) were assigned a higher priority, and those judged dependent (D) were assigned a lower priority. In some cases, the committee had to compare a chemical's global utility to ensure that it rated as dependent for any explosive preparations in which it could be put to use.

The committee sorted the chemicals into three groups based on whether they met the conditions of the higher priority for one, two, or three criteria. The committee placed chemicals that met the conditions of a higher priority for three criteria in Group A; for two criteria in Group B; and for one criterion in Group C. The final evaluation is provided in Table 2-2. Coincidently, the precursor chemicals sorted into three groups of almost equal size. In this study, the committee chose to conduct an in-depth examination of the Group A precursor chemicals.

The decision to include urea ammonium nitrate (UAN) solution in Group A represents the only departure from a strict application of the committee's ranking principles. UAN is considered a relatively new product with limited geographical distribution, but commercially available. There is a well-documented history of explosives production from analogous urea-nitrate salt solutions used in Iraq. While UAN has not been used historically to produce explosives, the ease of producing various explosives from nitrating urea solutions, as seen in Iraq, supports the notion of UAN as a future threat and justifies its inclusion in Group A.

There is an additional caveat for certain precursor chemicals insofar as they come in a diverse range of concentrations when contained in commercial products or bulk mixtures. For example, hydrogen peroxide as low as 35% can be quickly blended to make an explosive charge if mixed with the proper fuel. While some control strategies specify concentration thresholds (see Chapters 3 and 4), the lack of a scientific consensus on what those thresholds are precluded the committee from including concentration thresholds in the prioritized table (Table 2-2).

CONCLUSION

The National Academies' 1998 short list, which was later applied by the Department of Homeland Security (DHS) to construct the list of chemicals in the Chemical Facilities Anti-Terrorism Standards (CFATS) Appendix A, only focused on precursor chemicals that made charges with larger mass sizes suitable for VBIEDs. Looking at the trend in Table 2-1, more bombing incidents are reporting smaller charge mass sizes, consistent with PBIEDs. Based on this trend, the committee chose to cast a wider net, by looking at precursor chemicals that can be used to manufacture VBIEDs or PBIEDs, and further prioritized the precursors using three criteria: suitability for large and small charge sizes, hence VBIEDs and PBIEDs; prior use; and dependency.

Every chemical in Table 2-2 is viewed as a viable precursor chemical and a viable threat, whether it has been sorted into Group A, B, or C. Group ranking is

TABLE 2-2 Ranking of Precursor Chemicals into Three Groups

Group	Chemical	Charge Size	Prior Use	Dependency
Group A	Aluminum (powder, paste, flake)	V/P	Y	I
	Ammonium nitrate	V/P	Y	I
	Calcium ammonium nitrate	V/P	Y	I
	Hydrogen peroxide	V/P	Y	I
	Nitric acid	V/P	Y	I
	Nitromethane	V/P	Y	I
	Potassium chlorate	V/P	Y	I
	Potassium perchlorate	V/P	Y	I
	Sodium chlorate	V/P	Y	I
	Urea ammonium nitrate solution	V/P	N*	I
Group B	Calcium nitrate	V/P	N	I
	Hydrochloric acid	V/P	N	I
	Potassium nitrate	V/P	N	I
	Potassium permanganate	P	Y	I
	Sodium nitrate	V/P	N	I
	Sodium nitrite	P	Y	I
	Sulfur	V/P	N	I
	Sulfuric acid	V/P	Y	D
	Urea	V/P	Y	D
	Zinc (powder)	P	Y	I
Group C	Ammonium perchlorate	P	N	I
	Antimony trisulfide	P	N	I
	Hexamine	P	Y	D
	Magnalium (powder)	P	N	I
	Magnesium (powder)	P	N	I
	Pentaerythritol	P	Y	D
	Phenol	P	Y	D
	Potassium nitrite	P	N	I

NOTE: *See discussion for explanation of including UAN in Group A. V: VBIED, P: PBIED, Y: used historically, N: not used historically, I: independent, D: dependent.

somewhat subjective, and could change depending, for example, on the interpretation of existing data or a shift in terrorist tactics. Continuous reevaluation of the precursors is encouraged by the committee, as some of the rankings may change over time with an evolving threat environment. The committee concentrated its efforts on Group A chemicals when examining the supply chains and existing controls, both discussed in Chapter 3.

3

Domestic Chemical Supply Chain

This chapter characterizes the movement of precursor chemicals through domestic supply chains and the potential vulnerabilities inherent in those supply chains, based largely on the presentations provided during the data-gathering meetings. It does so by (1) mapping a supply chain, in general, for the precursor chemicals; (2) cataloging and overlaying existing domestic policy mechanisms that may improve the security of those precursor chemicals; and (3) singling out gaps in supply chain visibility and oversight. Of particular security concern are the possibilities of unexplained losses, diversion, theft, and other misappropriation of precursor chemicals at the nodes and in modes of transportation throughout the supply chain. Several policy terms used in this chapter are defined in Box 3-1.

In the course of describing the domestic policy landscape, the committee does not present evidence on the effectiveness of any particular policy mechanisms because the evidence is largely unavailable, but the committee does highlight some of the costs, including unintended consequences to businesses and users. The committee was able to obtain information on compliance and participation rates for some security programs, but not on the programs' contributions to risk reduction per se. The evidence might be missing because of the methodological challenges of discerning risk reduction (Appendix B) or the limited development of retrospective assessment.[43]

In this report, the committee does not discuss the quantities of precursor chemicals moving throughout the domestic supply chains in detail. That information would not provide insight on mechanisms to restrict access given that the amounts of precursor chemicals required to make a person-borne improvised explosive device (PBIED) are many orders of magnitude smaller than

> **BOX 3-1**
> **The Committee's Use of**
> **Mandatory, Voluntary, and Other Terms**
>
> Government officials, academics, and industry representatives each use the terms mandatory, voluntary, and regulatory differently to describe existing policy mechanisms, relationships between public and private entities, and the roles of government officials or industry representatives. In the statement of task for this report, the Department of Homeland Security (DHS) asks the committee to "suggest controls that could be considered as part of a voluntary or regulatory scheme," thereby implying the equivalence of the terms mandatory and regulatory. The committee recognizes the overlap of the terms *mandatory* and *regulatory*, but it does not use the terms interchangeably: mandates, at whatever level of government, often or even typically involve corresponding regulatory action, but can occur without regulation. Thus, the committee uses *mandatory* to describe policy mechanisms that are prescribed in and enforceable under federal, state, or local law or ordinance; *regulatory* to describe rules for implementation; and *voluntary* to describe policy mechanisms that lack the same legal basis and force as mandatory policy mechanisms.
>
> The committee further recognizes a gray area in distinguishing between mandatory and voluntary measures, largely involving industry-led initiatives and public-private partnerships. Under such arrangements, a business that markets precursor chemicals might choose to participate in a security-related program that requires it to adopt certain procedures, subject to penalty, in exchange for various benefits. Given that the company's initial decision to participate in the program was voluntary, the committee has chosen to treat those and similar arrangements as voluntary.
>
> For analytical clarity, the committee also draws a distinction between controls and other types of policy mechanisms. The committee reserves the term *control* for mandatory restrictions on access to precursor chemicals, and uses terms such as *measure* and *activity* more broadly, to encompass policy mechanisms, such as outreach, training, or reporting, that could be mandatory or voluntary, under different circumstances. Thus, a strategy for reducing access to precursor chemicals could include a mix of mandatory and voluntary policy mechanisms, with or without new controls.

the amounts contained throughout the commercial supply chain. For example, a representative of Yara International reported that 75 kilotons per year of calcium ammonium nitrate (CAN),[44] itself one of the less common Group A precursor chemicals, are imported into the United States, which is enough to construct about 12,000 Oklahoma City–size charges. Furthermore, in 2012, only 0.1 percent of the Pakistani CAN production was required to be smuggled into Afghanistan to meet the needs of insurgents.[45] Secondarily, while some trade groups can provide an estimate of the precursor chemical mass moved per annum, the ability to know where smaller quantities, outside of the initial bulk shipments, are being moved and consumed is beyond the scope of this committee. The committee does not discuss specific international entry points either.

The supply chains described in this chapter are specific to the present state of domestic commerce within the United States and do not necessarily apply to in-theater procurement activities in which bomb makers are currently engaged.[46,47] The committee received data on how groups can source materials from manufacturers around the world through legitimate channels, then move them from more stable adjacent countries to feed their activities.[48-50] Countering the flow of goods into these areas will require an international effort,[51] which is also beyond the scope of the committee's charge with respect to domestic access.

SUPPLY CHAIN OVERVIEW

Supply chains consist of multiple processes and activities, within and between companies, from planning and procurement, through the time when a product or material is transported, until it is delivered to the end user. Supply chains provide a virtual map of how commodities, raw materials, works in process, and finished goods move from origin to consumption or other end uses.[52] In the United States, supply chain transactions are well documented, providing visibility into who has possession of, who has specific responsibilities for, and who is the ultimate user of the product or material in question. This section provides an overview of general supply chains. Details about the supply chains of individual precursor chemicals in Group A can be found in Appendix D.

For example, a company might document the planning and forecasting processes leading up to placing a purchase order. The information used in these processes might include historical sales or consumption data and forward-looking estimates of customer needs based on marketing knowledge. Companies use the planning and forecasting processes to limit inventory and ensure adequate free cash flow for other needs. The forecast generally results in a purchase order that provides specific information regarding product quantities, specifications, point of origin, shipping destination, negotiated pricing, and expected delivery dates. In most organizations, this information is generated and stored electronically.

Subsequent documentation will include purchase order confirmations, the shipping mode, the carrier name, and bills of lading that may spell out specific requirements and possible regulatory responsibilities. Since most transactions are now tracked electronically, visibility of transfers and the movement of products and materials have improved dramatically in the last two decades. Just as consumers can track an online order from its origin to their home or office, so can companies track their orders electronically from purchase order to delivery, often with great confidence in the data.[53] While many companies focus on the benefits of improved visibility for inventory management, improved visibility may also make it easier to identify, track, and monitor suspicious product movements.

At the conclusion of any shipping process, a proof of delivery—physical or electronic—will be generated to verify that products or materials were received at a certain location in specific quantities.[54] In the case where products or materials are subsequently shipped to distributors, retail outlets, or consumers, additional documentation will be generated to indicate method of shipment, carrier, quantities, ship-to locations, and proof of delivery. Reports on in-transit incidents or accidents may also include information that can be helpful in reconciling total quantities of material when products are missing.

The most difficult part of the supply chain to document with surety is the final transaction to the end user. This is especially true if the transaction consists of unregulated materials or substances that fall below regulated thresholds (weight, volume, or concentration) or if the transaction is conducted with cash.

Some supply chains are very short or self-contained. For example, a company might manufacture a precursor chemical that it subsequently uses in an internal process, or ship a precursor chemical directly to an end user who then consumes it. The precursor chemicals might be used to manufacture additional, differentiated products for resale or might be applied in another process. In such cases, companies may provide their own internal transport by truck, rail, barge, or pipeline. Documentation for these moves, unless otherwise regulated, may be difficult to track.

Supply chains that extend beyond the boundaries of a single enterprise generally adhere to accepted practices for documentation and regulation. State and federal regulations apply to almost all shipments of goods, with special emphasis placed on materials that are caustic, toxic, or flammable, where weight and volume limits may be placed on shipments depending on the mode of shipment. This includes all supply chains that ultimately service or employ distributors, wholesalers, retailers, resellers, or consumers.

Throughout the study, the committee received data from industrial sources and trade groups on the supply, use, and consumption of precursor chemicals. Using this information, it constructed diagrams to illustrate how the chemicals move throughout the domestic supply chain (see Appendix D). Though the chains of the chemicals serve different industries and end users, there are many commonalities in the types of nodes that the precursor chemicals encounter as they move from origin to end use.

Figure 3-1 illustrates how precursor chemicals might move through a typical supply chain. Precursor chemicals enter the U.S. supply chain through imports or manufacturing operations and are subsequently transported by various shipping modes to points of use or to intermediate storage locations—such as distributers, wholesalers, or retailers—before being sold to customers. Each box in the illustration represents a node in the supply chain that constitutes a point of origin, mode of transportation, interim storage location, transfer of ownership, or end use. Chemicals enter the supply chain as precursor chemicals at blue nodes, are stored at gray nodes (excluding transportation), and are transformed

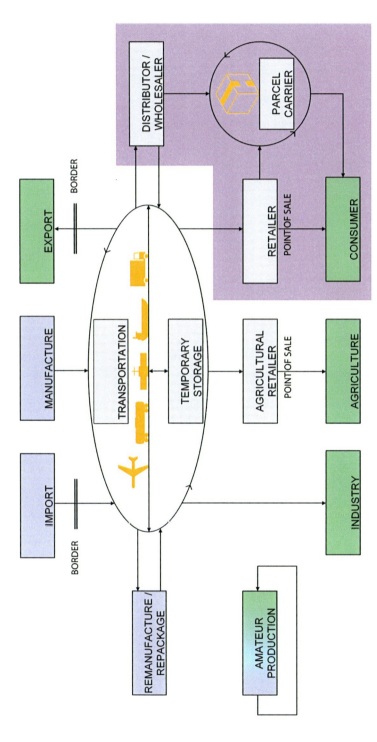

FIGURE 3-1 Generalized supply chain diagram showing the typical nodes encountered by precursor chemicals as they move from origin to consumption or other end use. Chemicals enter the supply chain at blue nodes, are stored at gray nodes (except transportation), and are transformed into something else or consumed directly at green nodes. Purple indicates the possibility of e-commerce.

into something chemically different or consumed directly at green nodes, while the nodes on a purple background indicate the possibility of e-commerce (see Appendix D for a full glossary of node designations). The following sections detail the types of nodes present on this diagram.

Production and Input Nodes

In Figure 3-1, blue nodes indicate where precursor chemicals enter the supply chain. These nodes include import, manufacture, and re-manufacture. Amateur production is unique in that it is a closed system. The green export node is discussed here as well.

Import/Export

The import and export nodes encompass transportation of the precursor chemicals into (import) and out of (export) the United States from or to foreign countries, either as bulk materials or finished products. The nodes are separated on the chart for conceptual clarity, but transactions in either direction will occur at the same physical locations. Specific locations may include seaports where large container ships offload their goods (e.g., CAN prills produced in Europe)[44] or land-based border crossings with Canada and Mexico (e.g., caustic or high-water-content precursor chemicals such as hydrogen peroxide).[55] Most ports also include terminal facilities for temporary storage or transloading; however, for clarity on the supply chain charts, these are treated as either part of the port node if covered by the Maritime Transportation Security Act (MTSA, see below) or as a variant of a commercial distributor if covered by the Chemical Facilities Anti-Terrorism Standards (CFATS).

Manufacture

The manufacture node refers to domestic locations where precursor chemicals are made from other raw materials via a chemical synthesis or industrial process. Such factories may either produce a precursor chemical directly or as a by-product of another process. An example of a direct synthesis would be the manufacture of nitromethane from hydrocarbons. Or, as seen with ammonium nitrate (AN) manufacturing, the excess nitric acid by-product is sold to other end users.

Remanufacture and Repackage

Unlike primary manufacture, remanufacture and repackage operations acquire bulk precursor chemicals and either incorporate them in further formulations or split them into smaller packages, with the requirement that the precursor chemical remains the identical chemical species before and after the reformulation. An example of remanufacture would be a match factory that uses potassium chlorate to make the match heads, while repackagers include locations that bag precursor chemicals as ingredients for fertilizer blends.

Amateur Production

The committee acknowledges that hobbyists and other members of the public can synthesize certain precursor chemicals themselves from readily available raw materials. An individual manufacturing precursor chemicals could ship or transfer them to another person through a carrier or other means; however, the committee does not have access to data to either confirm or deny that this occurs. To reflect the dual synthetic and consumptive role of these individuals, the node is colored blue and green. While instructional materials for at-home precursor synthesis are readily available, the committee does not list specific references or detail these protocols.

Transportation Modes

Transportation here means the loading, movement, storage incidental to movement, and unloading of property, including solid, liquid, and gaseous materials. The Group A precursor chemicals are either solids or liquid mixtures (e.g., some sodium chlorate and all urea ammonium nitrate (UAN) solutions are liquids during transport), and transportation is conducted primarily by vehicles. Shipment may occur in a wide variety of packaging formats, ranging in size from large tanks and hoppers for bulk materials to bottles and boxes for small quantities. During transport, the owner of the precursor chemical may have title, but not necessarily physical possession, in which case, the carrier might be obligated by contract or law to insure the owner against all or partial loss. A person who places a precursor chemical into transportation in commerce is defined as a shipper, whereas a person who performs the transportation function is a carrier, and both designations may apply to the same person. Carriers may be for hire—that is, a common carrier that transports property in commerce based on a fixed price for any person or entity—or private. A private carrier transports property it owns or for limited persons or entities under contract.

Ship and Barge

Precursor chemicals are imported and exported via ocean freight and may be transported domestically along inland waterways. In both cases, vessels move precursor chemicals in bulk as solids and powders in dry holds, sometimes bagged in bulk-size sacks, or as liquids in tankers.

Truck

Trucking is a major mode of transportation for both bulk material (e.g., hopper or tanker trucks) and small quantities and formulations of precursors (e.g., dry van and tractor trailers).

Rail

Rail is used primarily to transport bulk materials in tank cars, hopper cars, and boxcars. Certain precursor chemicals that are transported by rail require specialized means of containment because of their corrosive nature (e.g., nitric acid and hydrogen peroxide) or physical properties.

Air

Due to the high cost per volume of transport, only small quantities of precursor chemicals are shipped via air freight, mostly for specialty or individual use.[56] This shipping will fall primarily within a parcel carrier's distribution system.

Parcel Carrier

Shown separately in Figure 3-1, parcel carriers (e.g., USPS, UPS, FedEx) use trucks for the final delivery of small quantities of precursor chemicals to consumers. They employ their own distribution systems, separate from those of the transportation nexus, which may include truck, rail, and air shipping modes.

Pipeline

Within the United States, a several-million-mile-long pipeline network gathers, transports, and distributes gases and liquids to commercial and residential customers. The only case where a precursor chemical is transported via pipeline, outside of the relatively short pipelines within chemical plants, is in the distribution of UAN. Pipeline networks deliver that product to terminals for domestic distribution or for export.

Distribution and Retail Nodes

The gray boxes on Figure 3-1, other than transportation, are broadly grouped as locations where the precursor chemicals are physically stored and recorded for future sale and use, either in large or small quantities. Most of these locations differ from temporary storage locations insofar as title and responsibility have passed from the manufacturer or supplier to an entity that will ultimately distribute the product to consumers or other end users.

Temporary Storage

It is not uncommon for precursor chemicals to reside in temporary storage locations during transport (e.g., AN prills transferred to rail-side hoppers) for either short or extended periods of time while awaiting further movement or transloading.[57] These nodes are considered separate from distributors as they represent a transitory stopping point en route to a specific location.

Distributors and Wholesalers

Finished product may be sold to an intermediary that will hold the product in inventory, awaiting resale to other nodes. Distributors may work with specific products or with chemicals in general (chemical distributors), or may serve as warehouses and wholesalers for a variety of goods. The committee learned that some distributors also break bulk shipments of materials into smaller portions or allotments, for example, the bagging of AN fertilizers.

Agricultural Retailers

Agricultural retailers are considered a separate node because their customers and the uses of the precursor chemicals are highly specific. In this study, agricultural retailers are defined as local businesses that sell bulk quantities of agricultural chemicals to agricultural end users, as defined below, and store them onsite. The committee learned that many agricultural retailers provide application services to their customers, thus maintaining physical custody of the precursor chemicals until they are dispersed, and that a minority of these retailers bag products on request.[58]

Retailers

Products are sold to consumers, commercial and noncommercial, by entities designated on the chart as retailers. This node can assume a variety of forms, from physical home improvement stores and pharmacies to online storefronts and platforms.

End User Nodes

The green nodes in Figure 3-1 represent locations where precursor chemicals exit the domestic supply chain. This can be accomplished by export out of the domestic supply chain (see above), by chemical reaction transforming precursor chemical into a different species, or by the ultimate use of the chemical.

Industry

Industrial end users encompass manufacturers that convert a precursor chemical into another chemical or finished product, in which it is no longer the same chemical species. For example, one of the primary uses of hydrogen peroxide is in the pulp and paper industry as a bleaching agent, where it is consumed.[55] A finished product process might include cold casting aluminum powder, which transforms the powder into a solid piece of metal. Industrial uses not covered by this node are those that simply mix precursors without changing their chemical properties, such as the blending of nitromethane with furfural to make agricultural products; these users are considered remanufacturers or repackagers.[59]

Agriculture

This node is defined as commercial operations, ranging from family farms of a few acres to large-scale facilities, that grow food crops or other plant products. While farmers with landholdings across the size spectrum account for a significant portion of the node's end users, this node also includes other professionals such as landscapers and other horticulturists.

Consumer

In Figure 3-1, and throughout the report, a consumer is defined as a nonindustrial, nonagricultural end user who employs a precursor chemical directly for either commercial or noncommercial activity. The commercial category includes cosmeticians, who use peroxide-based bleaching products, and jewelers, who use nitric acid in metal finishing kits. Noncommercial uses by the general public, range from personal hygiene and home care to pyrotechnic and rocketry hobbies, but are limited to personal needs.[60]

Internet Commerce

E-commerce presents unique challenges to restricting access to precursor chemicals. The reach of the internet across municipal, state, and national borders can enable potential buyers to bypass local restrictions by purchasing from retailers in other jurisdictions, under different rules. Moreover, a buyer can remain anonymous by masking its identity behind multiple layers of obfuscating cover and using nonidentifying methods of payment or making purchases via the dark web, which is a venue for illegal activity.[61]

To the committee's knowledge, all the precursor chemicals in Group A can be purchased online and shipped to end users, including private individuals, from multiple sources. In some instances, retail sites automatically present buyers with purchase ideas, such as "People who bought chemical X also bought Y," thus suggesting, though not identifying it as such, an ingredient list for an explosive combination. That is not to say that a substantial share of internet sales of precursor chemicals lack legitimacy; for example, AN, nitromethane, and aluminum can be used in exploding targets, racing fuels, and pyrotechnics, respectively.

While online transactions can occur throughout the supply chain, this report limits the discussion of e-commerce to direct, consumer purchases from retailers or finished product manufacturers, delivered via a parcel carrier service, along with any transportation methods used to deliver those products. E-commerce sometimes blends node designations when the finished product manufacturers sell directly to consumers; however, for the purposes of the report, manufacturers retain their manufacturing designation, with the purple e-commerce zone and shipping arrows indicating consumer sales on the supply chain charts.

The committee learned that two types of e-commerce websites tend to sell precursor chemicals: retailers and platforms. Retailers are defined as companies that both provide the online storefront and fulfill orders. One class of retailers separated out on the supply chain diagrams (Appendix D) is the chemical supply companies that typically sell directly to research institutions or companies, with limited, sometimes vetted, sales to individuals.[62] More common retailers include internet-only retailers (e.g., Amazon) and companies that also operate at physical locations (e.g., Home Depot and CVS), both of which sell broadly, without vetting. Platforms differ from retailers insofar as the website merely facilitates transactions between other parties. For example, eBay and Craigslist connect buyers and sellers, but do not list or fulfill any product orders themselves. Any website allowing communication between parties could potentially serve as a platform for e-commerce of precursor chemicals.

The accessibility and usability of e-commerce has raised new challenges for restricting access to precursor chemicals. Through online channels, chemicals can flow directly to consumers, across national, state, and municipal borders with relatively little visibility, absent direct interventions to monitor or track information on retail listings, orders, and purchases. Without a physical presence and the opportunity to engage face-to-face with the intended purchaser, the retailer cannot, for example, identify suspicious behavior beyond the decision to purchase a particular chemical or combination of chemicals, which might not be suspicious on its own. What could be suspicious behavior might only manifest through direct engagement. Another challenge is the difficulty of tracking multiple purchases of the same or complementary chemicals from different sources, both brick and mortar and online.

Internet retailers, like most other retailers, are not currently required to restrict the sale of precursor chemicals; however, some have promulgated their own restrictions (e.g., Amazon bans certain Department of Transportation [DOT] hazmat classes,[63] and eBay maintains a list of banned chemicals[64]). However, these self-imposed restrictions do not always prevent the listing and sale of the targeted materials. For example, even if an internet retailer refuses to sell a material directly and does not allow others to sell it on the platform, other businesses or individuals who sell through the platform (e.g., non-Amazon retailers who sell through Amazon or eBay users) might be able to circumvent the policy.

eBay works with law enforcement agencies proactively and on request and automatically filters, removes, and prevents online listings of sales of prohibited items.[65] There are also reports of eBay-based transactions leading to the identification of potential suspects.[66] The committee requested conversations with other online retailers. Lacking a response, the committee does not know whether eBay's practices are industry-wide, or whether companies outside of the United States would work with domestic law enforcement agencies in any capacity.

DOMESTIC POLICY MECHANISMS

Federal controls that aim to halt bombing attacks have a long, if somewhat intermittent, history. In response to the United States' entry into the Great War, Congress passed the Federal Explosives Act of 1917, which dramatically curtailed access to both explosives and the "ingredients" (referred to as precursor chemicals in this report) used to make them.[67,68] This act was repealed after the war, then temporarily reinstated for the duration of World War II.

Presently, many policy mechanisms are in place throughout the domestic chemical supply chain that could directly or indirectly contribute to security objectives. Figure 3-2 shows that a wide range of controls and other policy mechanisms currently impact each node of the supply chain.

A policy mechanism, especially one that addresses security, can both reduce the probability that precursor chemicals will be used to make HMEs for use in IEDs and potentially mitigate the severity of improvised explosive device (IED) attacks should they occur (see Appendix B). Several of the mechanisms considered in this chapter are likely to impact probability more than severity. More specifically, they reduce the likelihood of IED attack consequences by securing the production, storage, or distribution of particular precursor chemicals.

The remainder of this chapter discusses the coverage of existing controls and other policy mechanisms to identify potential vulnerabilities. The sections discuss these controls and mechanisms, including private-sector initiatives, in greater detail. On the federal level, the controls—and the associated regulatory authorities—are sorted by department and agency. Consistent with the respective missions of each department or agency, the controls have different goals. For example, DOT's hazardous material regulations focus on safety; the primary interests of the Environmental Protection Agency (EPA) are public health and environmental protection; and the Department of Homeland Security (DHS) targets security with CFATS and MTSA.

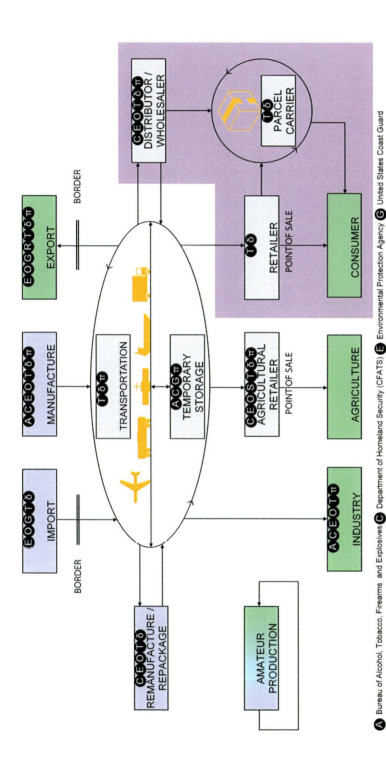

FIGURE 3-2 The generalized supply chain from Figure 3-1, now overlaid with relevant controls and other measures or activities (black circles).

Department of Justice

Attacks involving explosives are investigated and responded to by law enforcement at all levels of government. At the federal level, this falls under the authority of two Department of Justice (DOJ) bureaus, the Bureau of Alcohol, Tobacco, Firearms and Explosives (ATF) and the Federal Bureau of Investigation (FBI).[69] The latter is only concerned with incident investigations and counterterrorism operations, whereas the former also directly or indirectly regulates a subset of precursor chemicals and maintains the United States Bomb Data Center (USBDC), a national collection center for information on arson and explosives-related incidents throughout the country.

ATF directly regulates chemicals that are defined as explosive materials via a yearly updated list,[70] but only regulates one of the identified precursor chemicals (ammonium perchlorate less than 15 micron).[71] Nevertheless, ATF's regulatory oversight will apply at those nodes where the precursor chemicals are manufactured into explosives. These regulations inevitably leave exploitable gaps because they do not cover precursor chemicals per se or, as a related matter, the sale and transportation of unmixed products, such as binary exploding target kits (see below). The absence of certain materials from the ATF explosives list also may not represent the physical properties of some precursor chemicals, a prominent example being AN, which is capable of detonation in a neat state as has been demonstrated historically.[72-75] Although AN does not appear on the ATF explosives list, it is referenced as an "acceptor" subject to sympathetic detonation from the detonation of explosive materials stored nearby; AN is also considered a "donor" when it is stored within the sympathetic detonation distance of explosives or blasting agents, where distance is calculated using one-half the mass of AN to be included in the mass of the donor.[76]

A third DOJ agency that deals with precursors is the Drug Enforcement Administration (DEA), though for different purposes. DEA maintains lists of chemicals that are used in drug manufacturing, a potentially analogous activity to HME production.[77] A "List I" chemical is a chemical that, in addition to legitimate uses, is used to illegally manufacture a controlled substance and is important to the manufacture of the substance; a "List II" chemical is a chemical that, in addition to legitimate uses, is used to illegally manufacture a controlled substance, but without the designation of importance. There is no overlap between DEA's List I and the chemicals in Groups A, B, and C, but there is some overlap between the DEA's List II and the chemicals in Group B, namely hydrochloric acid, potassium permanganate, and sulfuric acid.[78] In addition, DEA's Special Surveillance List includes one Group C chemical, magnesium.[79] As with precursor chemicals used in explosives, drug precursors also have legitimate uses.[80] The process used by DEA to control drug precursors can provide examples of the potential challenges and results of regulating precursor chemicals in certain ways (Box 3-2).

BOX 3-2
Controlling Drug Precursors in the United States

Domestic controls on methamphetamine precursors provide an example of one approach to controlling precursor chemicals in the United States and of malicious actors' responses.

Starting in 1988, the federal Controlled Substance Act was amended to include controls on a list of precursor chemicals, including the methamphetamine precursor ephedrine. The framework of these controls shares many similarities with the control measures implemented internationally and at the state level in the United States, and includes provisions for record keeping, mass thresholds, requiring buyers to provide identification, and requiring sellers to report suspicious activity. However, a provision in the law exempted formulations approved by the U.S. Food and Drug Administration (FDA) as drugs, which allowed drug producers to still have access to ephedrine tablets. This exemption remained in effect until 1993,[81] at which time drug producers switched to ephedrine combination products and pseudoephedrine.

The full suite of methamphetamine precursors was added to the Drug Enforcement Administration's (DEA's) List I under the Comprehensive Methamphetamine Control Act of 1996.[82] While adding controls to pseudoephedrine, the target was primarily commercial sales, with over-the-counter transactions and blister packs not subject to oversight. As a result, pseudoephedrine continued to be a primary target of diversion.

In 2005, the Combat Methamphetamine Epidemic Act significantly expanded controls over pseudoephedrine in several areas.[83] The same record keeping, buyer identification, and suspicious activity reporting now apply to all transactions and are supported by a requirement for employee training in their implementation. The product is also controlled by requirements for keeping the product secured behind the counter. There are additional limits on the quantity that is available for purchase by a single person on a daily and monthly basis, and record keeping can be facilitated with national registries.[84]

DEA reports that, as a result of these laws, the amount of precursor chemicals in the domestic supply chain has decreased significantly, as is reflected in the reduction in the number of operational methamphetamine labs.[77] However, within that time frame, there has been a concomitant increase in trafficking from Mexico and small-batch domestic production to meet the same domestic demand for methamphetamine. Thus, while limiting access to the precursor chemicals might have decreased the production of methamphetamine in the United States, especially in labs, manufacturing has simply moved elsewhere or occurs by other means. Arguably, limiting access did not achieve the underlying goal of preventing the acquisition of the finished goods.

Department of Homeland Security

DHS is involved at several points along the precursor chemical supply chains through different offices and agencies: the United States Coast Guard (USCG) and Customs and Border Protection (CBP) are involved at import and export nodes; the National Protection and Programs Directorate (NPPD) applies security requirements to those facilities manufacturing or storing certain precursor chemicals; and the Transportation Security Administration (TSA) facilitates the credentialing of workers who move the products. Three DHS policy mechanisms are discussed in the next sections.

Chemical Facility Anti-Terrorism Standards

Initially authorized by Congress in 2007 and administered by NPPD, CFATS identifies high-risk chemical facilities via a screening procedure and regulates them to ensure that they maintain adequate security.[18,85-87] A high-risk facility is defined as one that contains certain listed chemicals at or above specified threshold quantities and concentrations. The list of relevant chemicals is published as CFATS Appendix A and sorted as posing risks of release (public health), sabotage (public life and health), or theft and diversion (weaponizable).[88] CFATS grants flexibility in the options for realizing adequate security, with DHS assisting the creation of workable security plans. Inspections to ensure compliance with the approved plans are concomitant.[86] DHS is authorized to use compliance orders, civil fines, and cease operations orders to enforce CFATS.

Like many prioritization lists, CFATS indicates threshold quantities and concentrations that determine when the regulation applies, with most Group A precursor chemicals listed at 400 lb as theft or diversion security risks, the exceptions being AN (2,000 lb) and aluminum powder (100 lb). About 3,000 of the evaluated facilities eliminated the use of the relevant chemicals or brought their inventory below the prescribed thresholds or concentrations; in the case of hydrogen peroxide, several end users and distributors began to request 34.5% solutions (with 35% as the concentration limit in CFATS).[55] There are also facility exemptions, including those covered by other regulations (e.g., MTSA) and, very specifically, water treatment facilities.[89]

A unique caveat is encountered at some ports that use terminals for transloading or temporary storage. If the terminal is within the grounds of the port, it will be covered by MTSA, while if it is not within that area it will be covered by CFATS. DHS presently does not plan to screen truck terminals for inclusion in the Section 550 regulatory program, and therefore DHS will not request owners and operators of truck terminals to complete the Top-Screen risk assessment methodology. For clarity, CFATS is not listed on the relevant supply chain diagrams at the port node, and the terminals are treated as a variant of a commercial distributor.

When constructing the CFATS Appendix A list, DHS encountered the same issue of ubiquity as described in Chapter 2. It reached the same conclusion: that prioritizing chemicals such as hydrogen peroxide and nitric acid instead of acetone and urea is the more logical choice. It cited the Academies' 1998 study to support that determination.[14]

The presence of a CFATS Appendix A chemical at a facility does not necessarily imply that it is regulated under CFATS. Of the 60,000 Top-Screen assessments submitted by 38,000 unique facilities,[90] only 2,570 are currently covered as high risk.[91] This is a point of concern, as those approximately 35,000 remaining facilities do not have to implement security plans, either because they did not meet the statutory requirements or because they took preemptive action to reduce their risk profile by reducing the quantities of precursor chemicals below mass or concentration thresholds. Because the CFATS thresholds were designed with the goal of preventing vehicle-borne improvised explosive devices (VBIED)s, the current methodology for assessing facility risk may not be well-equipped for preventing access to person-borne improvised explosive devices (PBIED)-relevant quantities of precursor chemicals.

Maritime Transportation Security Act of 2002

Managed by USCG, the goal of the MTSA is to prevent a Maritime Transportation Security Incident, defined as "any incident that results in: loss of life, environmental damage, transportation system disruption, or economic disruption to a particular area."[92,93] This is accomplished by establishing security procedures at all U.S. ports and vessels based on terrorism vulnerability assessments. Mitigation strategies generally include surveillance, security presence, and credentialing and identification materials. MTSA does not directly regulate the precursor chemicals, but dictates the security procedures and precautions that all cargo must be subject to at ports, thus securing them indirectly. In addition to MTSA, USCG follows hazardous materials transportation regulations proffered by DOT (see below),[94] while maintaining a sub-list of hazardous materials that require specialized handling procedures.[95]

Hazardous Material Transportation Credentials

Operators of trucks transporting hazardous materials, including most of the Group A precursor chemicals, are required to have a valid state-issued commercial driver's license (CDL) with a hazardous material endorsement. To acquire an endorsement, the CDL holder must undergo a threat assessment, including a background check and vetting against the Terrorist Screening Database, and fingerprinting.[96,97] MTSA requires a Transportation Workers Identification Credential (TWIC) for workers who need access to secure maritime facilities and vessels.[96] TSA conducts a similar security threat assessment to determine a worker's eligibility to receive a TWIC.

Environmental Protection Agency

EPA focuses on protecting public health by preventing the release of precursor chemicals that may affect the population. Therefore, EPA regulations do not directly address precursor chemical security from the standpoint of misappropriation.

Toxic Substances Control Act

Under the Toxic Substances Control Act (TSCA), EPA has broad authority to issue regulations designed to gather health, safety, and exposure information; require testing; and control exposures to chemical substances and mixtures at chemical manufacturers and importers.[98,99] TSCA gives EPA authority to take specific measures to assess the adverse health effects of new and listed chemical substances (>70,000) and to protect against unreasonable risks to human health and the environment from existing chemicals. Regulations may restrict or ban the manufacture, importation, processing, distribution, use, or disposal of any chemical substance that presents an unreasonable risk to human health or the environment. If the risk of a chemical substance is already managed effectively under a different statute, regulation under TSCA generally is not used.

All Group A chemicals except CAN and UAN are listed in TSCA, but regulation occurs only if the manufacturing facility exceeds 25,000 pounds per annum production. As of 2012, there are about 4,800 reporting facilities. CBP uses TSCA paperwork for imported chemicals before they are released from custody.[100]

Emergency Planning and Community Right-to-Know Act

The purpose of the Emergency Planning and Community Right-to-Know Act (EPCRA)—which is not, strictly speaking, a control—is to provide local governments, first responders, and the public with information on the potentially hazardous materials in their communities and to facilitate emergency planning in the event of material release.[101,102] This reporting may introduce an unintended security risk by providing malicious actors with information on the facilities that are storing precursor chemicals of interest.

Department of Labor

The Department of Labor (DOL) is concerned with the safety of workers who handle or are exposed in the workplace to chemicals, some of which are precursor chemicals that can be used to make explosives.

Occupational Safety and Health Administration

A subset of the precursor chemicals listed in this report are regulated directly by the Occupational Safety and Health Administration (OSHA) under two sections, either in relation to usage or storage at explosives-containing facilities (aluminum, AN, and chlorates)[103] or when defined as Highly Hazardous Chemicals.[104] These regulations focus primarily on the safety of the workers handling the materials, with only some storage requirements potentially impacting precursor chemical security, albeit indirectly.

Regardless of specific inclusion and contingencies, commercial nodes that handle precursor chemicals will be subject to OSHA's general workplace regulations. For example, most businesses are required to have emergency action plans (EAPs), which include contingencies for events such as a fire.[105] These EAPs require components, such as employee training, monitoring and alarm equipment, and evacuation plans. In some cases, close cooperation with local responders is required to maintain public safety. Because such regulations apply generally throughout the supply chain, individual supply chains (see Appendix D) only show symbols for those chemicals that are specifically considered highly hazardous by OSHA, as regulations are directly contingent on the presence of one of those precursor chemicals.

Mine Safety and Health Administration

For the purposes of this study, oversight by the Mine Safety and Health Administration (MSHA) only applies at end-user nodes that represent blasters that use commercial AN-based explosives where the precursor chemicals exist onsite in a neat, pre-mixed state. MSHA and ATF maintain an interagency memorandum of understanding regarding the enforcement of explosives regulations under such circumstances.[106] Given the limited scope of these regulations for the purposes of this report, MSHA is grouped implicitly with OSHA on the relevant charts.

Department of Transportation

DOT's primary interest with precursor chemicals is transportation safety, and it regulates all Group A chemicals except CAN (exempted under Special Provision 150) and UAN solution.[107,108] The Pipeline and Hazardous Materials Safety Administration (PHMSA) is responsible for regulating and ensuring the safe and secure movement of hazardous materials in commerce by all modes of transportation. The Office of Hazardous Materials Safety within PHMSA develops regulations and standards for the classifying, handling, and packaging of hazardous materials.[109] A system of placards and labeling of packages and a shipping paper—that is, a manifest—that contains the material's proper shipping name, class, division, United Nations identification number, and quantity as well as the number of packages, an emergency contact, and any special permits to which transportation is subject must accompany each shipment. Thus, the identification of hazardous materials cargo is readily available to enable first responders to address incidents during transport. During transport, drivers must adhere to approved routes (dictated by state and local governments) and cannot leave vehicles unattended, both contingencies providing a security benefit.[110] Security training is also required for those defined by regulation as hazardous materials employees. Both shippers and carriers of hazardous materials must obtain a Hazardous Materials Registration issued by PHMSA to perform their specific transportation functions.

Of the 520,000 truck carriers registered with the Federal Motor Carriers Safety Administration, 90,000 are authorized to transport hazardous materials.[110] Carriers of hazardous materials are required to maintain certain minimum levels of financial responsibility for the cargo they transport in commerce; for example, interstate commerce transporters of oxidizers such as AN are required to maintain a minimum of $1,000,000 of coverage compared to a minimum of $5,000,000 for any quantity of Class 1, 2, and 3 explosive materials transported in interstate commerce, which also requires security plans for any quantity transported and routing plans for transporting quantities in excess of 55 lb.[111,112] Some carriers are required to have a security plan for the transportation of a subclass of hazardous materials—including AN, hydrogen peroxide, nitric acid, and nitromethane—if they meet the specified threshold quantities.[113] Security plans may include requirements for transfers and attendance by personnel, depending on transportation method.[114]

Department of Commerce

The Department of Commerce's (DOC's) Harmonized Tariff Schedule lists all the prioritized precursors (in Chapters 28, 29, 31, 76, 79, and 81) and dictates the duty rates for each material's import.[115] DOC also lists certain chemicals as export restricted, Category 1 on the Commerce Control List.[116,117] In the cases of aluminum (1C111a.1) and magnesium (1C111a.2.a.3) powders, inclusion on the list depends on the average particle diameter of the material, for reasons of controlling missile technology, nuclear nonproliferation, regional stability, and antiterrorism. Equipment that produces aluminum powder is also restricted under 1B102c for missile technology and antiterrorism purposes. Magnalium (aluminum-magnesium alloy) powder is controlled under 1C1002c.1.d for nuclear nonproliferation, national security, and antiterrorism. AN formulations are controlled if meeting the definition under 1C997 (citing antiterrorism and regional stability) of containing "more than 15% by weight ammonium nitrate, except liquid fertilizers," which, while covering both AN and CAN, excludes UAN solution. A license is required for AN with respect to export or re-export to Iraq. Nitric acid is listed under 1C999e if greater than 20% for antiterrorism and regional stability purposes, with licenses required for North Korea and Iraq.

State and Local Regulations

In addition to federal oversight, the commercial movement of precursor chemicals may be subject to state and local laws or ordinances depending on the specific jurisdiction. These may be independent or in excess of existing federal regulations, or they may be close mimics. An example of the latter is that most states have incorporated DOT regulations on hazardous materials transport into their statutes. There are thousands of statutory jurisdictions within the United States, and the committee could not review all of their policies on precursor chemicals; however, at the state level, some existing regulations pertaining to AN provide an illustrative example of nonfederal controls.

In anticipation of DHS's AN rulemaking in 2008, several states implemented their own regulatory schemes that contain some common features (see Figure 3-3).[58,118-133] Of the fifteen states with such schemes, the most prevalent point of commonality is record keeping on retail-level AN transactions; these records typically contain the date, information on the identity of the purchaser, and the quantity sold. Eleven of the fifteen states with regulatory schemes also require that the retailer maintain a license to sell the material, but a buyer does not require a license to purchase it. Roughly half of the states also require storage

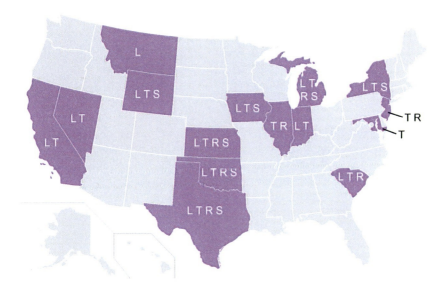

FIGURE 3-3 Coverage of state controls affecting AN via requirements for distributors and sellers to be licensed or registered (L); transaction details to be recorded (T); the refusal of sale when confronted with suspicious activity (R); and storage security (S).
NOTE: Indiana, Montana, Nevada, New Jersey, and South Carolina do not regulate AN specifically, but cover it under general regulations dealing with fertilizers.

security for AN and specifically allow retailers to refuse sales to people behaving suspiciously. The committee heard from a representative of the New York Department of Agriculture, who provided further details of enforcement, which includes liaising with law enforcement during inspections to help promote compliance. The representative noted that the number of outlets carrying AN in New York declined after the state introduced the regulation and discussed legal opportunities for farmers to obtain restricted chemicals from out-of-state and out-of-country distributors that might not apply similar protections.

In general, state and local provisions may apply at any node within a precursor chemical's supply chain, for example, through business permits, transportation route planning, or fire codes. Thus, given the thousands of political subdivisions in the United States and the committee's task not allowing for a complete analysis of all possible controls, these marks are generally excluded from the supply chain charts in Appendix D, but are assumed to be possible at all stages to some degree in some places. Only the state laws affecting AN are included in the overlays in Figure 3-2 and Figures D-1 through D-11.

The state AN laws illustrate some of the challenges of controlling access to precursor chemicals at the subnational level. With respect to coverage, even if a state has controls, malicious actors can easily cross state lines and obtain the material where it is not regulated; all the AN-controlling states have at least one such border. A second issue is the lack of harmonization between the states. When given the option, each state will implement controls differently, so while common themes can be found throughout Figure 3-3, the lack of a unified control scheme can engender confusion and adds additional burdens to commerce while creating situations where nonmalicious actors unwittingly violate the law.[134]

Private-Public Partnerships

For this study, private-public partnerships are defined as programs wherein the government engages with private entities on a voluntary basis, with the goal of providing additional benefits without statutory changes.[135] This can include programs that are directly managed by the government or those where statutory requirements may be accomplished by references to the literature produced by private entities.

Known Shipper Program

The Known Shipper Program (KSP) is a voluntary program originally created by the Federal Aviation Administration for strengthening air cargo security. The program establishes procedures for differentiating between shippers that are known and unknown for air carriers and indirect air carriers who tender cargo for air transportation.[136,137] A known shipper is a person who has an established business relationship with an indirect air carrier or operator based on records or other vetting. Currently, TSA allows using manual procedures, the Known Shipper Database, and the Known Shipper Management System to classify a shipper.

Customs-Trade Partnership Against Terrorism

The Customs-Trade Partnership Against Terrorism (C-TPAT) is a voluntary private-public partnership authorized by the Security and Accountability for Every Port Act of 2006 with the goal of improving commercial security.[138,139] C-TPAT partners work with CBP to develop security plans that protect their supply chains both from the introduction of contraband and from theft. In exchange, CBP allows for reduced inspections at the port of arrival, expedited processing at the border, and penalty mitigation. Compliance with the program is not free to shippers, and there are attendant costs.[139]

National Fire Protection Association

The National Fire Protection Association (NFPA) codes consolidate fundamental safeguards for the storage, use, and handling of hazardous materials in all occupancies and facilities (not necessarily with respect to security concerns), and are required by some levels of government and commercial operations. The codes do not apply to the storage or use of hazardous materials for individual use or residences. Relevant codes may include 400, Hazardous Materials (e.g., AN) or 484, Standard for Combustible Metals (e.g., aluminum powder). Due to the ubiquity of the NFPA codes and their adoption by myriad organizations, they are not shown on the supply chain charts.

Trade Associations Programs

Some trade associations condition membership on businesses' participation in additional safety and security programs. Secure handling improvements might be accomplished through technical measures that address safety or security, incentives related to employment and human resources, and social or behavioral pressures. Specifically, programs may set specific secure handling goals; promote awareness and process change; publicly recognize firm participation; and promote organizations that implement secure handling practices. The programs listed in the following sections are not exhaustive of all the efforts of trade associations that work with the precursor chemicals, but provide illustrative examples.

The reach of these programs has limitations. Even if trade associations condition membership on participation, not all businesses in a particular industry choose to join a trade association. The representatives of one trade association that provided data to the committee acknowledged that not all businesses opt for membership, despite the benefits that membership conveys. Thus, even if all members of a trade association participate in and comply with the association's program, vulnerabilities might remain.

Institute of Makers of Explosives

The Institute of Makers of Explosives (IME) provides members with best practice training materials via its Safety Library Publications and via a risk assessment tool (IME Safety Analysis for Risk) used to calculate risk to personnel from commercial explosives manufacturing and storage facilities and operations. The members of this trade association primarily deal with bulk precursors (e.g., unbagged AN), only work with companies possessing ATF licenses, and follow the specified safety guidelines published by the trade association. Compliance with these procedures is required for membership, and the trade association represents greater than 90 percent of the industry.[140] Publications relevant to this study include guides on AN and general transportation and storage.[141-143] The latter guides include, but are not limited to, controlled access to manufacturing locations, storage areas, and transportation containers through locks and seals.

International Air Transport Association

The International Air Transport Association (IATA) works closely with governments and the International Civil Aviation Organization and member airlines to develop regulations that advance safety and facilitate fast and efficient transport of dangerous goods by air.[144] The goal is to ensure that the regulations on dangerous goods transport are effective, efficient, and globally aligned. The IATA Dangerous Goods Regulations manual is a global reference for preparing, shipping, and transporting dangerous goods by air for the world's airlines. Participation in this group would mostly affect the small quantities of precursors shipped by air via parcel carriers.

American Chemistry Council

The American Chemistry Council's (ACC's) Responsible Care program has the stated goal of increasing the industrial performance of its members with regard to security and safety.[145,146] Areas of focus for security include site, supply chain, and cyber aspects. Participation in the program is a requirement of joining ACC and requires both performance reporting and third-party verification. The performance of specific companies is reported publicly to incentivize compliance (for a further discussion of voluntary compliance incentives, see Chapter 5). Similarly branded efforts have been used to build programs outside of the United States. It was reported that Responsible Care has significantly reduced health and safety incidents (53–78 percent depending on the measure) in member companies.

Society of Chemical Manufacturers and Affiliates

The Society of Chemical Manufacturers and Affiliates' (SOCMA's) ChemStewards, similar to Responsible Care, is focused on improving the health and safety practices of member companies.[147] Participation is a requirement for SOCMA members who manufacture and handle synthetic and organic chemicals. Training and verification are concomitant with the program.

National Association of Chemical Distributors

The National Association of Chemical Distributors' (NACD's) Responsible Distribution is a safety and security program using third-party verification of compliance.[148,149] Participation in the Responsible Distribution program is required for all members of the trade group, the bulk of which are small-scale distributor operations who may also break bulk shipments into smaller aliquots. Participation encompasses 250 distributors with a combined 1,900 facilities in the United States.

ResponsibleAg Inc.

ResponsibleAg Inc. is a nonprofit organization founded in 2014 to assist agribusinesses in complying with federal regulations regarding the safe handling and storage of fertilizer products.[58,150,151] The organization provides participating businesses with a compliance audit relating to the safe storage and handling of fertilizers, recommendations for corrective action where needed, and resources to assist in this regard. Program-credentialed, third-party auditors use an internal checklist to ascertain compliance every 3 years, and if the auditor identifies issues, the facility will receive a corrective action plan with recommendations; a facility is not certified until all issues are addressed. As of December 2016, 2,282 facilities have participated in the program, with 452 reaching full compliance.[150]

Outreach

Unlike the trade association–sponsored programs in the preceding section, those listed below are neither required for membership in any organization nor required by statute, and therefore might cover less of the relevant industries. Outreach efforts include both industry-led and government-sponsored activities, and may also include outreach from the federal government to state and local emergency responders.[152]

Government activities presented to the committee include efforts by ATF[153] and FBI[154] to increase awareness and reporting of suspicious activity. ATF has focused on educating and developing relationships with the fertilizer industry to limit illegitimate access to precursor chemicals such as AN by increasing voluntary reporting of suspicious activity involving the precursors and increasing awareness of security vulnerabilities. The program initiatives include *Be Aware for America, Be Secure for America,* and *America's Security Begins with You.* FBI has engaged in efforts targeting retailers (e.g., pool and spa and beautician suppliers carrying hydrogen peroxide). FBI reported that consistent results require ongoing outreach because of the high turnover rate of retailers.

Outreach in the form of sharing threat information with industry so that they can take appropriate precautions has also been reported. DHS has in the past provided information on federally sponsored training and education resources in which industry may partake.[155]

Best Practices

Best practices for security can make it harder to misappropriate precursor chemicals, and in this report are defined as those activities that are undertaken by an independent entity and are not part of any other mandatory or voluntary requirements. Best practices may include, but are not limited to, protection of assets by authentication, alarms, physical barriers, and facility personnel.[156] Some private industries also implement more stringent customer vetting to increase security, which may include background checks, inspections of receiving facilities, or global positioning system (GPS) tracking of shipments. There is also a set of publicly available literature on security practices, authored by both government[157,158] and private groups,[159] that offers more details for anyone who is interested in implementing best practices absent a more formal structure.

SUPPLY CHAIN VULNERABILITIES

If locks keep honest people honest by preventing unauthorized entry, then it could be said that laws, regulations, and voluntary measures will help ensure that honest, conscientious manufacturers, purveyors, and consumers of precursor chemicals oversee Group A chemicals adequately. However, the unexplained loss, diversion, or theft of precursor chemicals and the potential unintended consequences to the commercial enterprises that make or sell them cannot be taken lightly. The purpose of this section is to highlight potential areas of vulnerability in supply chains that handle precursor chemicals. The policy coverage shown in Figure 3-2 suggests that the potential for a malicious actor to acquire a precursor chemical increases toward the end of the supply chain.

Types of Vulnerabilities

Laws, regulations, and voluntary measures are in place primarily to ensure that precursor chemicals remain under the direction of legitimate possessors who handle and use the materials. The primary security concerns in the chemical supply chains are potential vulnerabilities to unexplained loss, diversion, and theft, any of which could lead to malicious actors having access to potentially dangerous materials. Additionally, given the widespread availability of the precursors at retail nodes, there is an opportunity for legal acquisition for illegitimate applications (e.g., by misusing exploding targets kits, described below).

Unexplained Loss

An unexplained loss, for the purpose of this report, is an amount of precursor chemical that disappears from the supply chain without knowledge of the cause. The cyclic counting of stock in inventory with a reconciliation of acquisitions and distributions is a method used to determine evidence of losses. Losses of certain hazardous materials that are precursor chemicals are not generally reportable to authorities unless there are state regulations requiring these actions or the precursor chemicals are CFATS chemicals of interest stored at CFATS-regulated chemical facilities that have approved security plans.

Unexplained losses may or may not signal a threat to security. If a precursor chemical is misplaced or lost due to inaccurate inventory record keeping, it may eventually turn up. Unexplained losses that remain unresolved should potentially be considered to be in the hands of malicious actors and therefore to pose a threat.

Diversion

Diversion, as defined by Stanton,[160] is "a form of misappropriation. It is an act of acquiring a product by means of deception. The types of deception vary and do not always include the failure to compensate the targeted company. In all cases of diversion, the common factor is that the targeted company causes the diverted chemical to be placed into transportation. Preventing diversion starts with knowing your customer." At regulated nodes of the supply chain, diversion would likely require a recognized or authorized buyer of a particular chemical to act as an intermediary for an unintended recipient. This action might jeopardize the legitimate buyer's credentials, if it needed credentials for the transaction and a regulatory agency recognized that diversion had occurred.

Theft

Theft, for the purposes of this report, is the misappropriation of precursor chemicals without permission or payment where the material is presumed stolen and missing. Theft may be internally reportable within a commercial enterprise, but may not be reportable—or reported—to any local, state, or federal law enforcement agency. A single large-scale theft of a precursor chemical might be difficult to execute because it would be noticed easily and raise alarms. Theft of small quantities of precursor chemicals—in one event or stretched over time— may not be detectable and could be mislabeled as losses.

Coverage of Controls and Other Policy Mechanisms

Importers, manufacturers, and remanufacturers tend to fall under greater scrutiny due to existing oversight, such as DHS's CFATS (see Figure 3-2 and Appendix D), that requires significant documentation and inspection of on-hand quantities of precursor chemicals. With some exceptions, namely industrial end users and agricultural retailers, visibility and oversight of precursor chemicals appear to decline as precursor chemicals make their way through the supply chain. Because of this, the potential for misappropriation becomes more apparent at the later stages in the supply chain.

It is conceivable that the physical transfer of chemicals from one entity to another, for example, involving transportation or temporary storage nodes, represents an opportunity for thieves to gain direct access to the chemicals. Safeguards are generally in place in the transportation modes through verification of receiving weights or weigh station records, but they may not prevent malicious actors from extracting small quantities of chemicals. Although distributors, wholesalers, and retailers are cognizant of the threat of internal theft in warehouses and distribution centers,[161] as well as in retail outlets open to the public, the quantities of missing material might be small enough to avoid concern or suspicion.

A brief inspection of Figure 3-2 suggests that there are few controls or other policy mechanisms in place at retail outlets, apart from those in the agricultural sector. Parcel carriers also lack coverage, but might benefit from substantial, inherent traceability.

Of particular note, the committee found a pronounced lack of visibility and oversight in retail-level, nonagricultural transactions, especially those involving e-commerce, suggesting ample opportunity for malicious actors to acquire precursor chemicals for making HMEs. In many cases, consumers can legally purchase or acquire precursor chemicals, either in raw form or as components of other products, for legitimate commercial and noncommercial purposes. Hobbyists who manufacture amateur pyrotechnics for personal use might work with aluminum powder and potassium chlorate to create a display, and individuals who operate radio-controlled aircraft or cars might use nitromethane as fuel. Apart from its agricultural uses, AN is also an ingredient in some widely available household products. Many cold packs that are commonly sold in grocery and drug stores for use in first aid contain AN, as do exploding target kits used by archers and sporting firearms enthusiasts (see below). In the case of cold packs, it would not be difficult to remove the active chemical and put it to a different use.

The legal, retail acquisition of precursor chemicals for nonagricultural purposes might often occur in small quantities, either in person or online, but determined malicious actors could conceivably acquire a significant amount of material in a relatively short period of time. They might, for example, make multiple purchases, across multiple outlets, using cash or other nonidentifying purchase arrangements to avoid identification.

Moreover, as the Oklahoma City and Oslo bombings have demonstrated,[23,24] malicious actors can also pose as commercial entities to purchase large quantities of precursor chemicals without verification. In the United States, subnational restrictions and voluntary programs, such as Responsible Ag, have emerged over time that might reduce the risks of similar actions today in many domestic jurisdictions, but there are no federal controls in place that require retailers to ask for identification, reason for purchase, or other means of verification.

EXPLODING TARGETS

While every precursor chemical in Groups A, B, and C can be legally acquired by noncommercial users as both neat and finished products (see Appendix D), not all of the formulations pose the same threat of misuse for making HMEs. Commercially available binary exploding target (ET) kits provide an easy route of precursor chemical acquisition in the United States. Typically, the ET kit will have either AN or potassium perchlorate as the oxidizer and powdered aluminum as the fuel, all three of which are Group A chemicals. The kit's components are not explosive materials until mixed.

At the time of this report, most commercially available ET kits use AN—likely given its easier and safer handling—while potassium perchlorate is used in specialty ETs that require heightened sensitivity, to be initiated by less-energetic rounds. The capabilities of ETs are well described and readily apparent on the internet, which has given them high visibility to anyone searching for ways to produce IEDs. States and federal law enforcement agencies, such as FBI, have expressed concerns over the potential use of ETs in IEDs.[162] Because the AN-containing kits are a common, complete finished product available to the general public—and because they have been implicated as components in recent, high-profile attacks in New York and New Jersey (Table 2-1)—their properties are detailed below.

Chemical Characteristics

The danger of ET kits is of greater concern relative to other precursor sources because the precursor chemicals contained in them have been designed to produce a high explosive that is both detonable and requires minimized amounts of energy to be initiated. Additionally, unlike other products containing chemical precursors, no technical knowledge or expertise in chemistry is required to use the kits to produce an HME—indeed, they are packaged with instructions on how to use the components to make an HME—thus removing any barrier of use for would-be malicious actors. While the kits are highly accessible, people can and do construct their own ETs from other precursor sources.

Form

The two components, AN and aluminum, exist in a variety of forms, with different porosities and particle sizes used in different applications. Particles with higher surface area to volume ratios maximize contact between the fuel and oxidizer and increase the rate of reaction, and thus the explosive power. By including the two components in the optimal physical form for making an explosive, ET kits remove the need for malicious actors to process or refine the precursor chemicals to make a usable HME.

Weight Ratio

To make the most easily initiated binary explosive, the oxidizer and fuel components must be the proper particle sizes and be mixed in the proper proportions to ensure detonation (see the Detonability section below). In ET kits, each component is packed in the precise physical form and ratios so that when mixed the optimum conditions are created to ensure easy initiation of the HME from the impact of a bullet.

Detonability

Many HMEs require a detonator to initiate the charge. Many AN mixtures in particular require not only a detonator, but also contact with a larger mass of explosive set off by the detonator to ensure that the HME functions. ETs do not require a detonator as they are sensitive to lower shock inputs (e.g., a bullet impact). This property lowers the difficulty of deploying ETs in IEDs.

Legal Considerations

Federal and state regulations that control AN and aluminum also apply to ET kits, provided the packaged precursor chemicals meet the thresholds stated in the regulations. Even when these regulations do not control the ET kits or their components, however, federal and state entities may issue guidance on ET kit acquisition and use because of the potential for their misuse to make HMEs for IEDs.

Federal

The federal government does not directly regulate the sale or use of ETs, except for the U.S. Forest Service, which bans their use on federal land in the Rocky Mountain region to prevent forest fires.[163] While ATF and DHS regulate the commerce and security of explosives and precursor chemicals, respectively, as described previously, neither has the mandate or authority to regulate ETs.

ATF does not regulate the distribution and sale of ET kits because the individual components, when unmixed, do not meet the definition of explosive materials that establishes ATF's regulatory jurisdiction.[164,165] An ATF manufacturing license is not required by the consumer to mix the components because mixing falls under personal use. Transportation and storage requirements do apply once the components are mixed.

DHS oversees precursor chemicals for security reasons. However, unless the quantity of ETs located at a particular facility brings the total weight of AN or aluminum within CFATS thresholds (2,000 and 100 lb, respectively), those locations are not subject to DHS oversight. Additionally, if those thresholds were met, DHS would only regulate the security at the location, not transactions.

Similar to the outreach efforts by ATF to agricultural retailers and by FBI to other retailers, FBI has released a private-sector advisory for retailers that carry ET kits. This advisory describes different types of suspicious behavior, describes steps that retailers can take if they believe a transaction is suspicious, and provides a phone number for reporting that activity. Some ET kit manufacturers include FBI advisory, relevant ATF statutes, and prohibitions of use on certain federal lands on their websites. They also include instructions for persons to check for compliance with state and local laws.[166]

State

In lieu of federal controls, a number of states address ET kits at the retail or consumer level of the supply chain. As in the case of state regulations that control AN (see Figure 3-3), the regulations vary by state (see Figure 3-4).

Two states have laws under consideration that would directly address ETs: Rhode Island[167] would ban ETs outright, whereas New York[168] would require a valid permit for purchase.

Five states currently regulate the purchase of ET kits via licensing or permitting, either explicitly or implicitly. Washington[169] requires a license and California[170] requires a permit to lawfully possess the components with the intent to combine them to form an HME. Massachusetts similarly bans the possession of materials that can be combined to make a "destructive device" without "lawful authority."[171] Maryland[172] and Louisiana[173] have instead approached the issue by redefining the term "explosives" in their relevant statutes explicitly to include two or more nonexplosive materials that are sold together and that when mixed become explosive. There is some ambiguity in Virginia due to their explosives statute, where the interpretation of the exemption for "lawful purpose" may or may not include ETs.[174] The Attorney General of Virginia issued an opinion in 2014 on the issue, stating that possessing and using ETs for their designated purpose (i.e., recreational use) is legal.[175]

Ten states regulate ETs under statutes that require a license to manufacture and possess explosives.[176-185] In these states, purchasing the ET kit is legal, but mixing the components is illegal without a license. In New Mexico, the manufacture and possession of ET kits for "legitimate and lawful . . . sporting purposes"

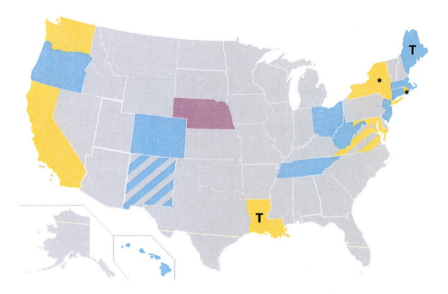

FIGURE 3-4 State regulations for ETs and ET kits. The state requires a license or permit to purchase or possess an ET kit (yellow), mix the components of an ET kit (blue), or use an ET (magenta). Stripes indicate ambiguity in the statute regarding lawful use. States with proposed laws are marked with an asterisk (*), and those with a 5 lb licensing threshold are marked T.

is lawful, depending on the interpretation of the statute.[177] To the committee's knowledge, authorities in at least four states—Ohio (State Fire Marshal),[186] Tennessee (Attorney General),[187] Connecticut (Office of Legislative Research),[188] and Massachusetts (State Fire Marshal)[189]—have rendered the opinion that, when mixed, exploding targets are subject to existing statutes on explosives. Tennessee requires a license to manufacture explosives but does not issue such licenses, thus effectively banning the mixing and use, but not the purchase, of ET kits within the state.

Nebraska falls into its own regulatory category because it only regulates the use of explosives, not the possession of them or their components.[190] Thus, it is legal to purchase ET kits and mix the components, but detonating the ET requires a permit.

Two states provide the additional caveat that the possession and use of less than or equal to 5 lb of unmixed (Louisiana[173]) or mixed (Maine[185]) ET products does not require a license.

CONCLUSION

This chapter highlights the reduced visibility and oversight as precursor chemicals flow through the supply chain and approach consumers. Often, this decreasing oversight is driven by the amount, weight, or volume of materials on hand at a given node. As illustrated in Table 3-1, many chemicals have thresholds (quantities and/or concentrations) below which oversight and inspections are no longer required. Businesses might maintain precursor chemicals below the threshold to avoid regulation, as with the example cited previously of industry reducing the concentration of hydrogen peroxide in products to fall below the CFATS threshold of 35%.[55] Generally, the longer the supply chain and the smaller the deliverable quantity, the less visibility there is of the material.

For most of the precursor chemicals, the total number of entities and affected parties for each type of node greatly proliferates as the precursor chemicals move down the supply chain. Returning to the AN example, only six domestic plants manufacture AN, but thousands of agricultural retailers and farmers use the material, and even more nonagricultural entities or individuals use AN-containing products. In this respect, the opportunities for misappropriation of precursor chemicals increase with increasing distance from the point of origin in the supply chain. Moreover, enterprises that depend on their trade in chemicals and need their licenses or permits to stay in business might be more aware of risks and take greater precautions to secure facilities, guard inventories, maintain records, obey regulations, and conform to agency oversight requests than those dealing with smaller quantities that represent a small fraction of their primary business.

Regulation and oversight decline as quantities or concentrations fall below thresholds; however, sufficient small quantities can be amassed to cause concern. These small quantities can be obtained illegally or legally, but are all available legally.

The committee speculates that the likelihood of unexplained loss, diversion, or theft is inversely proportional to the amount of material being sold or stored. In other words, the smaller the quantity of product that changes hands, the less likely it is that a control or other policy mechanism governs the transfer, making it easier to misappropriate. Because there is little federal, state, or local oversight of small-scale transactions, the most likely point of illegitimate acquisition of precursor chemicals is at retail-level nodes, especially nonagricultural or noncommercial (Figure 3-5).

Regarding legitimate sales, purchasing small amounts of many of the precursor chemicals can go unnoticed because there is no requirement to track or log the sale and no assurance of a payment trail. While larger transfers of material might trigger an electronic record, small cash-like exchanges might remain anonymous. Consumers can obtain all Group A precursor chemicals online from multiple sources, and enforcing current, voluntary restrictions on internet sales presents challenges. Moreover, the European Union, discussed in the next chapter, suggests that mandatory restrictions on e-commerce can also present challenges.[191]

TABLE 3-1 Summary of Controls and Thresholds

		ATF	CFATS	EPA	USCG	OSHA	DOT	DOC	DEA
Group A	Aluminum Powder		100	X^T		X^B	$X^‡$	I, EX (<200 μm)	
	Ammonium Nitrate		2,000 (≥33%)	$X^{D,T}$		X^B	$X^†$	I, EX (≥15%)	
	CAN			$X^{D,T}$				I, EX (≥15%$_{AN}$)	
	Hydrogen Peroxide		400 (≥35%)	1,000^T (>52%)		7,500 (≥52%)	$X^†$ (≥8%)	I	
	Nitric Acid		400 (≥68%)	1,000^T	X (<70%)	500 (≥94.5%)	$X^†$	I, EX (≥20%)	
	Nitromethane		400	X^T		2,500	$X^†$	I	
	Potassium Chlorate		400			X^B	$X^†$	I	
	Potassium Perchlorate		400				$X^†$	I	
	Sodium Chlorate		400		X (<50%$_a$)	X^B	$X^†$	I	
	UAN Solution			X^D	X			I	
Group B	Calcium Nitrate			$X^{D,T}$			X (NCF)	I	
	Hydrochloric Acid		15,000 (≥37%)	5,000^T (≥37%)	X	5,000 (0%$_W$)	X	I	X
	Potassium Nitrate		400	$X^{D,T}$			X	I	
	Potassium Permanganate		400	100^T			$X^†$	I	X

Group B	Sodium Nitrate		400		X	I
	Sodium Nitrite		100^T		X	I
	Sulfur		X^T	X (M)	X (M,P)	I
	Sulfuric Acid		1,000	X	X†	I
				1,000 (65-80%)		
	Urea		X^T			I
	Zinc Powder		1,000^T		X‡	I
Group C	Ammonium Perchlorate	X (<15 μm)	400		X†	I
	Antimony Trisulfide		X^T		X† (>0.5%$_{As}$)	I
	Hexamine		X^T			I
	Magnalium Powder		X^T		X‡	I, EX
	Magnesium Powder		100		X‡	I, EX (<60 μm)	X
	Pentaerythritol		X^T			I
	Phenol		500/ 1,000^T	X	X†	I
	Potassium Nitrite				X†	I

NOTE: A: aqueous; B: only when stored at a blasting location (under 1910.109); D: only when in an aqueous solution (under N511); EX: export restricted with caveats in parentheses; I: listed on the Harmorized Tariff Schedule; M: molten; NCF: not certain fertilizers; P: powder; PPF: powder, paste, and flake; T: TSCA threshold of 25,000 lb/yr; W: water; X: listed without threshold mass. A DOT security plan is required if transported in excess of 3,000 kg or L ($\dagger$), or if other conditions are met ($\ddagger$), but does not apply to 4.2 aluminum powder, <20% nitric acid, non-fuming sulfuric acid, or farmers.

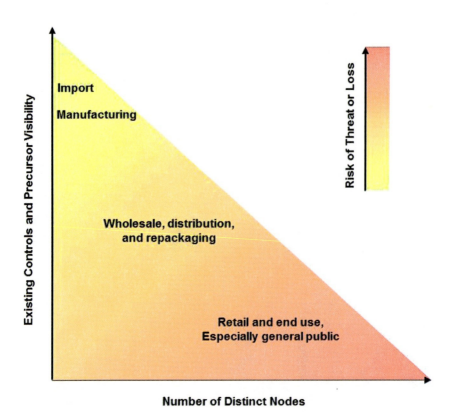

FIGURE 3-5 A visualization of the supply chain nodes in terms of oversight and vulnerability.

This chapter shows that the lack of uniformity of controls and mechanisms at the national and subnational levels has resulted in a regulatory environment that ranges from permissive to highly restrictive, depending on the precursor chemical, the supply chain node, and the location of purchase. Differences in states' rules create motivation for transactions across state lines, resulting in purchases by citizens that conflict with their state's regulations. The Joint Program Office for Countering IEDs, established in 2009 by FBI, was created to improve interagency coordination, reduce duplication of efforts, and systematically eliminate gaps in security that could be exploited by terrorists or other criminals. Further coordination among federal, state, and local agencies, and across independent and overlapping governmental and nongovernmental programs, can strengthen policy implementation.

4

International Regulations

Other countries have preceded the United States in developing policies to address concerns about access to precursor chemicals, and their experience can inform decisions about policy, including control strategies, in the United States. Table 4-1 compares regulations on precursor chemicals across the globe. Authorities generally agree on the types of chemicals that should be controlled, but differ in how they enact and implement policy. This chapter summarizes the policies of Australia, Canada, Singapore, the European Union (EU), and the United Kingdom.

To supplement publicly available documentation on policy and programs, two members of the committee and one staff member visited the European Commission (EC) in Brussels, Belgium, and the United Kingdom's Home Office in London, England, to learn firsthand about how they regulate precursor chemicals; the committee developed a list of questions that it sent in advance of these meetings to guide the discussions (see Appendix E). Information gathered from these meetings is described in the EU section.

The countries were selected because of their collaborative relationships with the United States; however, each has taken different mandatory and voluntary approaches to restricting malicious actors' access to precursor chemicals. The committee investigated the EU and United Kingdom's programs because of the challenges faced implementing new regulations and voluntary measures across Europe using a common framework, but allowing different strategies in each of the EU member states (MS). The committee examined Canada's program because of Canada's proximity to the United States and its streamlined approach to counter-terrorism. Australia provides an example of a whole-of-government-and-industry approach that is implemented within a decentralized framework.

TABLE 4-1 Comparative Chart of Global Regulations on Precursor Chemicals

	Chapter 2	CFATS	Australia	Canada	EU	Singapore	PGS
Acetone					X		X
Aluminum powder	A	X			X		X
Ammonium nitrate	A	X	≥45%$_s$	>80%	>16%$_N$	≥80%	X
Ammonium perchlorate	C	X	≥65%$_s$ 10%$_a$	X		X	
Antinomy trisulfide	C						
Barium nitrate						X	
Calcium ammonium nitrate	A				X		X
Calcium nitrate	B				X		
Guanidine nitrate						X	
Hexamine	C				X		
Hydrochloric acid	B						
Hydrogen peroxide	A	≥35%	≤65%	>30%	>12%	≥20%	X
Magnalium powder	C				X		
Magnesium powder	C	X			X		
Magnesium nitrate hexahydrate					X		
Nitric acid	A	≥68%	≥30%	>68%	>3%		X
Nitrobenzene		X					
Nitromethane	A	X	≥10%	X	>30%		X
Pentaerythritol	C						
Perchloric acid			X			X	

Chemical	Group					
Phenol	C					
Phosphorus		X	X			
Potassium chlorate	A	X	≥65%$_s$, 10%$_a$	X	>40%	X
Potassium nitrate	E	X	≥65%$_s$, 10%$_a$	X	X	≥5%$_a$
Potassium nitrite	C					≥5%$_a$
Potassium perchlorate	A	X	≥65%$_s$, 10%$_a$	X	>40%	X
Potassium permanganate	B	X				
Sodium azide		X	>95%			
Sodium chlorate	A	X	≥65%$_s$, 10%$_a$	X	>40%	X
Sodium nitrate	B	X		X	X	≥5%$_a$
Sodium nitrite	B					≥5%$_a$
Sodium perchlorate			≥65%$_s$, 10%$_a$		>40%	X
Sulfur	B		X			
Sulfuric acid	B			X		
Tetranitromethane					X	
Urea	B					
Urea ammonium nitrate solution	A					X
Zinc powder	B					

NOTE: %$_s$: percent composition in a solid mixture; %$_a$: percent composition in an aqueous solution; the letters in the second column correspond to the prioritized groups from Chapter 2; CFATS: Chemical Facilities Anti-Terrorism Standards; PGS: Programme Global Shield; the EU column includes both Annex I and Annex II chemicals.[192]

AUSTRALIA

After numerous incidents—the 2002 Bali bombings, which killed dozens of Australians; the conviction of five Sydney men in 2009 for a terrorism-related conspiracy; and a 2010 incident in which a homemade explosive (HME) exploded prematurely during transport, accidentally killing two men in Adelaide—the Australian government, along with state and territory governments, businesses, and industry, developed a voluntary National Code of Practice for Chemicals of Security Concern, which was enacted in July 2013.[193-195] The Council of Australian Governments (COAG) identified ninety-six chemicals of concern and developed a voluntary code of practice for fifteen high-priority precursor chemicals that can be used to make HMEs (see Table 4-1).

The goals of this code are to protect against theft and diversion, promote precursor chemical awareness between business and law enforcement, and educate personnel on suspicious transactions. While there is no direct federal government involvement since it is a voluntary code, the government does provide guidelines for implementation of the code in several areas, including the following:

- security risk management, which assesses the security risk of each business and assigns a security management point of contact who is responsible for investigating and reporting security incidents to the National Security Hotline;
- security measures chosen to reduce the risk of terrorists acquiring chemicals from each business by providing background checks on potential employees, educating staff on suspicious behavior, maintaining inventory control records, and reporting thefts or diversions; and
- supply chain security to verify the identity of all customers and promote transportation security of chemicals.

Security sensitive ammonium nitrate (SSAN) is subject to state and territory regulations and is not subject to the national code.[196,197] In June 2006, the COAG introduced a licensing regime for the use, manufacture, storage, transport, import, and export of SSAN, which comprises pure ammonium nitrate (AN), AN mixtures containing greater than 45% AN, and AN emulsions; SSAN does not include solutions or Class 1 explosives (explosive substances and articles, and pyrotechnics). The SSAN regulations form part of each Australian state's explosives regulations. Transporters moving more than 20 kg of SSAN or storing more than 3 kg need an annual background check. They undergo inspections to ensure that they have secure, lockable storage; maintain detailed records of usage; and have a security plan.

Unfortunately, it is not clear how successful this voluntary code alone has been on mitigating terrorist acquisition of precursor chemicals because of other counter-terrorism–related laws that were passed around the same time, including a telecommunications act[198] that allows companies to keep a limited amount of metadata for 2 years; an act against foreign fighters[199] that allows Australia to arrest, monitor, investigate, and prosecute as well as cancel passports for returning foreign fighters; and a counter-terrorism act[200] that responds to operational priorities identified by law enforcement and intelligence and defense agencies. Australia has not had a major terrorist attack on its soil. The attack by Numan Haider on two police officers in September 2014 and the siege at Martin Place in December 2014 have been the only successful terrorist attacks in Australia, and neither involved precursor chemicals.[201] Additionally, focused counter-terrorism efforts conducted by intelligence and law enforcement agencies have resulted in the apprehension of terrorists plotting events in Australia.[202-206]

CANADA

Natural Resources Canada's Explosives Safety and Security Branch administers the Explosives Act and Explosives Regulations for the safe and secure handling of explosives.[207] The Explosives Act

- requires anyone working with explosives to have a license, certificate, or permit issued by the Minister of Natural Resources;
- makes exceptions to this requirement for some explosives and storage activities and for the use of certain low-hazard explosives, low-hazard pyrotechnic devices, sporting ammunition, and consumer fireworks; and
- covers fireworks, pyrotechnics, propellant powders, ammunition, rocket motors, and restricted components.

Part 20 of the Explosives Regulations discusses the sale of and security plans for restricted components. The restricted components are the chemicals listed in Table 4-1 under the Canada heading. Parties must apply to sell a restricted component or to manufacture a product containing a restricted component. The application includes the applicant's personal information, the restricted components to be sold, the address of each location where a restricted component will be stored or sold and the storage capacity, and the name and personal information of the contact person for each location where a restricted component will be stored or sold. If AN is to be sold, the applicant must also submit a security plan and notify purchasers and/or transporters of security requirements, in addition to producing annual inventory reports that are delivered to the Explosives Regulatory Division (ERD) Chief Inspector of Explosives. The purchaser of any restricted component must provide photo identification (ID) and state the intent of use.

Since the implementation of this regulation, the ERD has seen an increase in products approved for sale in Canada and in importation permits, and has increased its compliance monitoring activities. The ERD developed a rating system for assessing the number and types of deficiencies found during inspections. It noted a decrease in reported deficiencies as a result of its compliance monitoring and assessments.[208] It is unclear whether the Explosives Act has had any impact on terrorism in Canada. Since 2009, Canada has charged twenty-one individuals with terrorism-related offenses that have been associated with bomb-making activities.[209]

SINGAPORE

Singapore has been targeted by terrorist groups and remains at a relatively high security alert. It has partnered with other nations in counter-terrorism.[210-212] It has also conducted major counter-terrorism–related exercises to assess and refine its readiness for terrorist attacks.[213] While there have not been terrorist attacks in Singapore, the country has developed proactive measures at its borders to enhance and improve government agencies' internal cooperation regarding precursor chemicals.

In 2007, fifteen hazardous substances listed in the Environmental Pollution Control Act and Regulations were delisted and instead listed in the Arms and Explosives Act.[214,215] The majority of the hazardous substances appear on the prioritized list of precursor chemicals in this report (Table 4-1). The Arms and Explosives Act clearly details the concentrations at which a material comes under its regulation. Singapore implements a licensing program to import, export, possess, manufacture, deal in, or store any of these fifteen substances. Possession is prohibited without a license. Application for a license can be made online or via phone. The license is precursor specific, and a licensee must undergo a background check and establish experience with and knowledge of how to handle the precursor chemical. All transactions must be recorded.

EUROPEAN UNION

The EU passed Regulation 98/2013 on the use and sale of explosives precursors in September 2014.[28] The policy action introduced a common regulatory framework across Europe regarding the "making available, introduction, possession and use, of certain substances or mixtures that could be used in the manufacture of home-made explosives." The regulation restricts access to and use of seven restricted explosives precursors (listed in Annex I of the regulations) by members of the general public. EU MS may grant access to these substances to the public through a system of licenses and registration, but a ban is the default. In addition, the regulation introduces rules for retailers who place such substances on the market. Retailers must ensure the appropriate labeling of restricted precursor chemicals and must report any suspicious transactions involving either the restricted or other nonrestricted substances that are considered items of concern (the latter listed in Annex II).

Under the regulation, each of the EU MS designates a competent authority and implements the ban, licensing, or registration processes. The authority is required to set up its own national contact point(s) for the reporting of suspicious transactions. The EC is tasked with providing the list of measures and guidelines to facilitate implementation.

The Standing Committee on Precursors

The Standing Committee on Precursors (SCP) is an expert group composed of representatives of EU MS and industry associations. It was established under the 2008 EU Action Plan on Enhancing the Security of Explosives and is chaired by the EC.[216,217] As depicted by the EC, the SCP provides a platform for EU MS and representatives of the operators in the supply chain to exchange information and share lessons learned on implementation.[191] It also helps facilitate the implementation of Regulation 98/2013 with the goal of limiting the general public's access to precursor chemicals and encouraging suspicious transactions and appropriate reporting of significant disappearances and thefts throughout the supply chain, while seeking minimal market disruption.

To create awareness in support of these goals, the SCP has issued guidance materials to inform the authorities and retailers of the EU MS of their roles and responsibilities under the regulation (see Appendix F).

Labeling, which is required, is another tool that can raise awareness but might not enhance security. For example, labeling might inadvertently alert bomb makers that a product can be used to make bombs. If the purpose of labeling is to provide retailers with information on targeted products so that they can meet their legal requirements, then retailers need to be able to get the information on products from manufacturers and/or suppliers; thus, notification in shipping materials might be as useful as labels, without jeopardizing security. The questions of the intended target of labeling (i.e., the customer or the retailer) and the purpose of labeling have not been answered.[218]

Compliance Effectiveness

EU Member States

Regulation 98/2013 defaults to a ban unless the EU MS tailor the regulation to meet their specific circumstances (i.e., licensing and/or registration). According to the EC's January 2017 report,[191] most EU MS were in compliance; the report included the following highlights:

- all EU MS identified one or more points of contact for suspicious transaction reporting and thefts;
- twenty-three EU MS were in full compliance, and had set rules and penalties, disseminated the implementation guidelines, and notified the EC of exceptions that would require licensing or registration; and
- five EU MS were in partial compliance, having not developed rules or penalties.

To promote compliance, the EC established regular bilateral discussions focused on implementation and potential regulation compliance issues with some EU MS.[219] An updated list of measures is available via the EC, which details each member's choice with regard to regulatory strategy (Figure 4-1).[220]

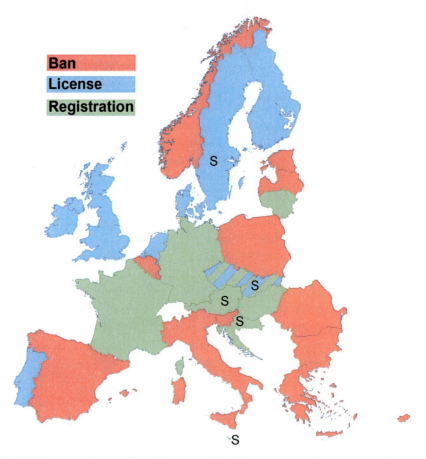

FIGURE 4-1 Implementation of Regulation 98/2013 by country. S: the country licenses or registers a subset of the precursor list.

While some EU MS have opted for the default, a ban on public access, others have chosen combinations of licenses and registration. There is significant variation in the processes for requesting licenses, including the criteria on which the requests are evaluated for approval or refusal, and the length and type of validity. Some EU MS accept that the request for a license should be granted unless there is a specific reason to refuse, while others refuse licenses unless there is a specific reason to grant them. Rates of licensing thus differ greatly from state to state, and there are no known instances of mutually recognized licenses.

The EC also reports that some EU MS have adopted additional measures that, for example, require retailers to register with the authorities and to periodically declare all transactions, including imports; extend the scope of the regulation to cover professional users; determine conditions for storage; foresee the exchange of relevant cross-border information with other EU MS; or establish a role for customs authorities.[191] Additionally, most EU MS have reportedly conducted awareness-raising campaigns that target the retailers marketing the precursor chemicals. Campaigns aim to increase awareness about the obligation to both restrict access and report suspicious transactions. Some EU MS have actively engaged online suppliers and marketplaces.

The United Kingdom

The United Kingdom's Control of Explosives Precursors Regulations 2014, which applies to England, Scotland, and Wales (Northern Ireland is covered separately[221]), came into force on September 2, 2014, as a means to implement Regulation 98/2013, and provides a case study on the approach of a particular EU MS to implementation. The United Kingdom's Home Office summarized the regulations as follows:

- the public may obtain a 3-year license to obtain and use the seven regulated precursor chemicals above their concentration thresholds, provided the applicants pass a background and medical check;
- licenses may be issued subject to conditions, for example, storage, use, quantities, concentration, and reporting loss or thefts;
- the supplier must verify the license and associated photographic identification of the purchaser and record the transaction details on the back of the license, and suppliers of Annex I substances must also ensure that such substances are labeled as restricted; and
- any violation of the regulation is punishable with up to 2 years imprisonment or a fine or both, with the possibility of lesser penalties for lesser offenses, such as minor clerical omissions.

As part of the regulatory process, the United Kingdom's Home Office, in cooperation with the Ministry of Justice and Her Majesty's Treasury, developed an impact assessment, much like a regulatory assessment in the United States (see Appendix G). The assessment considered the benefits, costs, and risks associated with doing nothing, or implementing a ban, licensing, or a registry in relation to three policy objectives:

- prevent terrorists from using precursor chemicals to make explosives;
- provide a mechanism for alerting authorities of illicit activities; and
- minimize the burdens on industry and legitimate users.

With that assessment, the United Kingdom chose to implement a licensing scheme, but not an accompanying registration program. Meetings with stakeholders suggested that an auxiliary registration program would create unnecessary confusion.[222]

Regulation Effectiveness

The EC reports[223] that Regulation 98/2013 has contributed to reducing the threat posed by precursor chemicals in Europe based on findings from meetings and consultations of the SCP and a study carried out by an independent expert consortium:

- **The amount of precursor chemicals available on the market has decreased.** This is partly because many retailers are applying the restrictions or have opted to stop selling the chemicals and partly because, voluntarily, some manufacturers have stopped making them. The supply chain has not reported any significant disturbances or economic losses as a result of this. Also, in some EU MS that maintain licensing, authorities have reported that the number of license applications is currently significantly lower than it was during the first year of application of the regulation. This suggests that members of the general public have successfully adopted alternative (nonsensitive) substances for continuing with their legitimate nonprofessional activities.
- **The capacity of law enforcement and designated authorities to investigate suspicious incidents involving precursor chemicals has increased.** EU MS have reported an increase in the number of reported suspicious transactions, disappearances, and thefts due to greater awareness among retailers who handle precursor chemicals. In addition, some EU MS have, on an ad hoc basis, exchanged information on reports and refused licenses. Finally, the authorities in EU MS that maintain licensing regimes have a better understanding of which members of the general public are in possession of restricted substances and the purpose they intend to use them for.

The EC has noted that it is not yet possible to assess the impact of the regulation on terrorist activities in more detail; however, EC officials indicated during the meeting in Belgium that some EU MS have thwarted potential terrorist activities, suggesting that the application of the regulation has contributed to the EC efforts to prevent terrorist attacks involving HMEs.[223] The EC also reported several challenges and costs, some of which are outlined below.

Challenges and Initial Responses

Preliminary evidence suggests that Regulation 98/2013 is contributing to security.[191] For example, in the United Kingdom, successes in reporting dangerous activities are available via a government-sponsored podcast,[224] and certain individuals who reportedly provided materials to terrorists have been identified.[225] However, there are challenges and costs incurred in several areas of implementation:

- authorities have had difficulty reaching all suppliers to inform them of their duties;
- authorities have had difficulty enforcing the restrictions and controls on internet sales, imports, and intra-EU movements, especially of small quantities;
- retailers have had difficulty identifying products that fall under the scope of the regulation;
- labeling, which is required, conveys risks of inadvertent disclosure;
- businesses that operate across intra-EU borders face the challenge of compliance with multiple regulatory schemes;
- economic operators have encountered difficulty determining whether the purchaser is a professional user or a member of the general public, because they do not have a process for establishing and verifying qualifications (one unnamed country is allegedly developing standardized industrial classification codes of those professions likely to need access); and
- evolving security threats present a continuous challenge that requires regulatory adaptability.

For a complete presentation of these challenges and some responses, see Box 4-1.

GLOBAL SHIELD

In an effort to mitigate the improvised explosive device (IED) threat, the World Customs Organization, in conjunction with Interpol, the UN Office on Drugs and Crime, and the U.S. Immigration and Customs Enforcement, initiated a 6-month pilot program, Programme Global Shield (PGS), in November 2010.[226]

> **BOX 4-1**
> **Challenges Arising from the**
> **Implementation of Regulation 98/2013**
>
> The EC reports a number of challenges relating to the implementation of Regulation 98/2013 and some initial responses. This box reproduces the relevant text of that report in its entirety.[191]
>
> > The main challenge for Member State competent authorities is the large number of operators affected by the restrictions and controls of the Regulation. Because many of the chemical substances and mixtures concerned by the Regulation are household products, the supply chain is significantly larger than that of other products subject to specific control provisions (e.g. drug precursors). Consequently, it has been challenging for competent authorities to reach all economic operators in the supply chain of explosives precursors to inform them of their duties. However, competent authorities have, in collaboration with the associations that represent the chemical industry and retail sector, conducted awareness-raising campaigns and engaged with a wide range of operators—from manufacturers to retailers, big companies to small independent stores, and internet sellers to marketplaces.
> >
> > Another challenge for Member State authorities is to enforce the restrictions and controls on internet sales, imports and intra-EU movements. Products having a large volume are often transported and sold in greater quantities (e.g. fertilizers), and are therefore relatively easier to identify and control. Conversely, other products are sold in small quantities and volumes, and are thus more difficult to intercept when they are shipped or transported into and across Europe. To address this problem, law enforcement and customs authorities are increasing their efforts to identify cases of illegal acquisition and possession by, for example, increasing inter-agency cooperation and exchanging information at EU level.
> >
> > A main challenge for economic operators, particularly those in the retail sector, has been to identify products that fall under the scope of the Regulation. Products containing restricted explosives precursors must be labelled accordingly. When that is not done early on in the supply chain, it is difficult for operators at retail level to properly verify

The program was aimed at enhancing awareness and information sharing on the global movements of fourteen chemicals (see Table 4-1).[227,228]

Program operations included training and establishing national contact points with the responsibility of providing monthly shipping reports on these fourteen precursors. A centralized multilingual web-based communication channel was set up to report suspicious movements and to act as a repository for project communications.[229]

that the label is affixed and that the restriction applies. In addition, the explosives precursors listed in Annex II of the Regulation that are not restricted do not have to be labelled. Economic operators, particularly those with high staff turnover, need to devote considerable time resources to identifying products of concern and training their staff appropriately.

Economic operators that conduct business across intra-EU borders also face the challenge of having to adapt to the specific nature of the different regimes in each Member State. The Regulation allows Member State authorities to define key aspects of its application in their territory. Consequently, economic operators must be aware of the type of regime that applies in the specific Member State the product is destined for, and must register the sale, verify a license or ban the sale, accordingly. Some companies have robust due diligence internal procedures which facilitate compliance with complex regulatory frameworks. However, for companies without such procedures, often smaller companies, this is a time-consuming process.

Finally, economic operators who sell to both members of the general public and other type of end-users have reported difficulty in assessing, with a reasonable degree of certainty, whether the person acquiring a restricted substance is acting for purposes connected with his trade, business or profession, or otherwise. Indeed, the Regulation does not specify criteria for evaluating what qualifies as professional or for checking professional credentials. These considerations are down to Member State authorities and, as a result, criteria may differ across different territories.

A continuous challenge for the Commission is to keep abreast of the evolving security threat. In order to adapt the Regulation to the use of new chemical substances, or of new concentrations of the listed substances, the Commission relies on information and data from member states. In 2016, efforts channeled through the SCP have led to three threat substances being added to Annex II (see Annex II, Section 4).

The World Customs Organization endorsed the program as a long-term endeavor, with the overall objectives of promoting international cooperation to prevent precursor chemical diversion, interdicting illegal shipments, and promoting awareness among industry and other groups.

Between 2011 and 2014, PGS resulted in the seizure of 40–100 tons of precursor chemicals per year.[230] This increased dramatically in 2015 to about 544 tons of solid precursor chemicals, largely due to sustained seizures along the

Afghanistan-Pakistan border by local agencies. The bulk of the amount (about 487 tons) consisted of AN and urea. In addition to precursor chemicals, PGS resulted in the seizure of detonators and completed HMEs and IEDs.

CONCLUSION

This chapter shows a range of approaches to restricting access to precursor chemicals and highlights the resulting challenges, some of which might pertain to a U.S. strategy and therefore merit further consideration. While a variety of approaches are used, the effectiveness of the programs is difficult to assess and compare, let alone quantify. Different avenues of terrorist attack will not be equally likely across countries given the different laws and regulations (e.g., the control of firearms in the United States versus the United Kingdom) and threat environments in each country. Thus, for the purposes of controlling precursor chemicals, it is necessary to gather data on the international controls only as they relate to the goal of limiting access to the precursor chemicals, and not their ability to deal with terrorism as a whole.

Market Level

Policy in other countries tends to focus on retail-level points of sale, whereas policy in the United States tends to focus on manufacturing, storage, and distribution. The EU defaults to a ban but allows flexibility for EU MS to implement licensing or registration for sales to the general public. Canada and Singapore have implemented similar licensing or registration measures for controlling access to precursor chemicals. By contrast, as described in Chapter 3, policy in the United States rarely addresses retail facilities or transactions, except in the case of agriculture.

Preliminary evidence of the retail-level controls implemented in the EU indicates a decrease in the amount of precursor chemicals that are available on the market, and an increase in the reporting of suspicious transactions, but with some costs to commerce and legitimate users.

Responsible Entities

Another difference between most of the countries described above and the United States is that precursor chemicals in the United States are not regulated by one responsible entity. This is similar to Australia, where there is a national guidance, but authority is devolved throughout its federal system.

In the United States, many federal and other agencies are involved to varying degrees along the supply chains, including the Department of Homeland Security (DHS), the Department of Transportation (DOT), the Environmental Protection Agency (EPA), and the Occupational Safety and Health Administration (OSHA).

Each maintains its own regulations and programs, intended to meet respective objectives for security, safety, the environment, and health. Because of this configuration, the committee found inconsistencies and possible gaps in coverage.

Harmonization

The United States might face similar challenges as the EU in its efforts to harmonize policy across EU MS. As the foregoing presentation of challenges and responses suggests, differences in implementation across MS might make it difficult to enforce restrictions on internet sales, imports, and intra-EU trade, and might require costly adaptations. With open internal borders, the EU is only as secure as its most vulnerable point of purchase. If malicious actors cannot obtain a precursor chemical in one member state, they can try another. In the United States, differences between federal law, state law, and local ordinance, and related concerns about the role of government, could present analogous challenges, depending on how policy unfolds going forward. The authoring committee of the 1998 Academies report observed that "a criminal in the United States can easily purchase precursor chemicals in a state with few requirements and then use them anywhere in the United States."[14] Furthermore, malicious actors now have the option to purchase precursor chemicals online, another widely available means of interstate trade.

EU regulations, PGS efforts, and HME collaborations between and among countries illustrate global cooperation in developing methods to minimize the broad access that terrorists have to precursor chemicals. These efforts as well as country-specific efforts will continue to increase compliance and awareness with current law enforcement security programs. While many countries conduct separate counter-terrorism response drills, cross-border drills can be conducted to understand the flow of information and resources and the changes in custody of precursor chemicals.

5

Assessing Possible Control Strategies

The preceding chapters present the results of the committee's deliberations on chemicals of interest (Chapter 2), domestic supply chains (Chapter 3), and the status of domestic and international policy (Chapters 3 and 4, respectively). The results uncovered potential vulnerabilities, particularly at the retail level of the domestic supply chain, and suggest opportunities both to reduce the likelihood that malicious actors will obtain precursor chemicals and to gather information to prevent or respond to improvised explosive device (IED) attacks, without undermining legitimate commerce and use.

This chapter focuses on tradeoffs among possible control strategies, derived largely from domestic and international experience, for mitigating vulnerabilities and gathering information. The committee did not presume the necessity of new controls, defined as mandatory restrictions on access to precursor chemicals (see Chapter 1 for definitions), or prejudge the legal standing of other policy mechanisms. Instead, the committee explored three types of strategy with new controls and one type without. In addition, the committee considered other measures and activities, such as outreach, training, and reporting, that could operate under public- or private-sector auspices, with or without legal mandates. These other measures and activities could serve supplemental or independent roles in different contexts.

More specifically, the committee considered the economic and noneconomic benefits and costs of the strategies, in relation to three policy objectives:

- restricting malicious actors' access to precursor chemicals;
- gathering and disseminating information to prevent or respond to terrorist episodes; and
- minimizing the burdens on legitimate commerce and use.

Each benefit or cost, be it economic or noneconomic, represents a factor for comparison in the Department of Homeland Security's (DHS's) statement of task (Box 1-1).

To assess the tradeoffs among control strategies, the committee engaged in an analytical exercise in which it walked through the strategies one by one and discussed the benefits and costs of each, qualitatively. It drew from evidence collected in the project and from principles of regulatory assessment, set out by the Office of Management and Budget (OMB) in Circular A4,[231] which provides guidance to federal agencies on the development of regulatory analysis (see Appendix G). For example, OMB specifies that a good regulatory analysis should include a statement of the need for the proposed regulatory action, an examination of alternative approaches to addressing the need, and an evaluation of the benefits and costs—quantitative, if possible, and qualitative otherwise—of the proposed action and the main alternatives.

OMB sets out a hierarchy of analytical methods, preferring benefit-cost analysis (BCA), which, ideally, would monetize all the benefits and costs of a proposed regulatory action, over other methods (see Appendix G). Regulatory analysis, even in the ideal, which the committee did not attain, cannot address all aspects of the tradeoffs among control strategies, but the underlying premise of each method—that policy interventions entail benefits and costs that can differ by approach—suggests means to systematically consider the relative merits of different strategies and inform the selection of one strategy over another.

One might describe the exercise as a notional analysis of benefits and costs and a starting point for a more detailed and rigorous analysis of policy options, but not as a BCA per se. Noteworthy among the deviations from a conventional BCA are the following: the committee did not attempt to identify all the benefits or costs associated with each control strategy; the committee lacked the data and resources to monetize or otherwise quantify the benefits and costs that it did identify; and the committee framed most of the benefits as pathways to attaining policy objectives (e.g., the capability to restrict access to precursor chemicals, the capability to track and correlate suspicious behavior), not as actual policy outcomes, such as reductions in bombings, injuries, or deaths.

Nevertheless, the committee was able to rank each of the three strategies with a new control based on the committee's assessment of individual benefit and cost attributes. For those three strategies, the committee was able to say whether a particular benefit or cost, exclusive of various uncertainties, was expected to be higher or lower under that strategy than under one of the others. The committee could not say how much higher or lower a benefit or cost might be from one strategy to the next, or whether one benefit or cost would outweigh another under the same strategy, but it was able to use the rankings to draw insight into tradeoffs among the strategies.

For purposes of this exercise, the committee set out the broad contours of all four strategies, but left many of the specifics that would be necessary for a complete qualitatively or quantitatively oriented policy analysis to future efforts.

POSSIBLE CONTROL STRATEGIES

Potential vulnerabilities, as shown previously, exist primarily at retail-level points of sale, especially nonagricultural. Although gaps in visibility and policy coverage exist elsewhere in the supply chains, the committee chose to focus in this chapter on retail-level sales as the highest priority.

A strategy to mitigate potential vulnerabilities could include different combinations of new and existing controls and other measures and activities, such as outreach, training, and reporting, which could be mandatory, voluntary, or a mix. At one end of the spectrum, all such measures and activities could be mandatory and, at the other, the measures and activities could all be voluntary. The committee presents the measures and activities without specifically designating them as mandatory or voluntary and comments later in this chapter on the implications of choosing a mandatory or voluntary route.

Retail-Level Controls

The committee's review of the policy landscape (see Chapters 3 and 4) suggests four general types of control strategy for retail-level transactions; three include a new mandatory restriction on access to precursor chemicals (i.e., a control) and one does not. The three strategies with new controls would implement one of the following at the retail level: a ban, licensing, or a registry. The fourth strategy would augment federal, state, and local controls that exist at the time of this report with other measures and activities not currently in place. The committee refers to the fourth strategy as "business as usual plus" (BAU+). Any of the three new controls could also be implemented in conjunction with other measures and activities.

A retail-level ban, licensing, or registry could apply to all or a just subset of retail-level sales, with the subset depending, for example, on the type of transaction or product market. Given concerns about the potential for commercial disruption, the committee chose to focus on applications of those controls to noncommercial sales and exempt commercial transactions, but policy makers could include commercial sales within any of the frameworks. A ban, licensing, or a registry could also include a restriction on quantities of purchases, as with the pseudoephedrine controls aimed to limit illicit drug manufacturing (Chapter 3).

Each control would likely require regulatory action, which could be prescribed at either the federal, state, or local level, with due consideration of preferences for state or local autonomy, differences in circumstances among states and localities, and the need for coordination and harmonization across state and local borders.

A ban on noncommercial sales might require a purchaser's evidence of commercial status to demonstrate their exemption from the ban; licensing for noncommercial sales might be implemented by requiring either evidence of commercial status, to demonstrate exemption from licensing, or a license to purchase; and a registry for noncommercial sales might be implemented by requiring either evidence of commercial status, to demonstrate exemption from the registry, or participation in a registry, for example, with a signature and government-issued ID. In each case, the requirement to present evidence—of commercial status, a license to purchase, or a signature and government-issued ID—might fall on the purchaser, but the seller might also be required to validate credentials. For example, a retailer might be expected to check the expiration date of a license or ID and to be able to recognize a counterfeit.

BAU+ would not add any new retail-level control, but it would not preclude action on other measures or activities, discussed below. Moreover, any strategy, even BAU+, could include a provision that grants retailers the right to refuse suspicious sales, but this might require additional legal or regulatory action.

Other Retail-Level Measures and Activities

Different combinations of other measures, involving either legal mandates or voluntary participation, could accompany a ban, licensing, a registry, or BAU+:

- training of retailers to, for example, request and verify evidence of commercial status, licenses, or government-issued IDs; identify suspicious behavior, fraud, theft, or loss; or fulfill responsibilities for reporting and documentation;
- reporting of suspicious behavior and fraud, theft, or loss to federal authorities and local law enforcement, respectively; and
- documentation of transactions, which could involve electronic record keeping and data analytics.

In each case, the measure could be publicly or privately sponsored. Other activities could include outreach for awareness, random audits, and systematic inspections.

The federal government, several state governments, and trade associations already conduct some outreach on precursor chemicals, their regulation, and safety (Chapter 3), but they and others could do more to educate retailers on precursors, including the requirements associated with any new controls. In addition, government officials, trade associations, or other third-party verifiers could evaluate retailers' implementation of new controls or other measures through random audits, systematic inspections, or a combination of the two. Random audits would catch retailers off-guard, whereas systematic inspections could allow time for self-assessment and preparation.

Some EU MS, including the United Kingdom, employ mystery shopping, a practice by which government officials—or potentially others—attempt to purchase regulated precursor chemicals incognito and then provide feedback to retailers on the interaction and whether it met regulatory requirements or aligned with guidance. In the EU, the approach serves more as an educational tool than as an enforcement tool, but it can also support enforcement objectives. Although targeting retailers, mystery shopping can also inform government program managers; for example, some mystery shoppers have uncovered opportunities for agencies to improve their outreach. For the purposes of this report, mystery shopping is considered a subset of audits and inspections.

Product placement might also play a part in each strategy. Retailers could choose or be required to place products that contain precursor chemicals behind the sales counter or in a locked case, so that consumers must request them directly. The extra layer of interaction and engagement might provide retailers with an opportunity to assess whether the purchase is suspicious and should be reported or denied, if they have the right to refuse a sale.

Building a Control Strategy from Controls, Measures, and Activities

The committee assembled three control strategies as packages of policy mechanisms, with each strategy featuring either a ban on noncommercial sales, licensing for noncommercial sales, or a registry for noncommercial sales, together with an unspecified limit on the quantity of noncommercial sales and a right to refuse sales to anyone, under suspicious conditions. Each control strategy also included a suite of supplemental measures and activities, consisting of outreach, training, reporting, documentation, auditing, and inspections. To facilitate the comparison of strategies, the committee held the measures and activities constant across strategies. Each strategy treats commercial sales identically—for example, commercial purchasers must provide evidence of their commercial status to complete a transaction—but treats noncommercial sales differently.

The committee also looked separately at a BAU+ strategy that included additional outreach, training, reporting, documentation, auditing, and inspections, absent any new control. Appendix H includes a summary of findings on cautionary labeling as well, but the committee did not include cautionary labeling in any of the strategies because of concerns raised in the EU and elsewhere about unintentional knowledge transfer (Appendix H).

Figure 5-1 illustrates the development of a strategy from a menu of options, starting with the selection of a new control, if any, and proceeding to the selection of measures and activities that can either supplement the new control or accompany existing controls. In the example, which mirrors one of the strategies that the committee considered, the resulting package consists of licensing with training, reporting, documentation, outreach, auditing, and inspections.

The committee did not specify whether a particular measure or activity would be voluntary or mandatory, but envisioned that, in either case, retailers

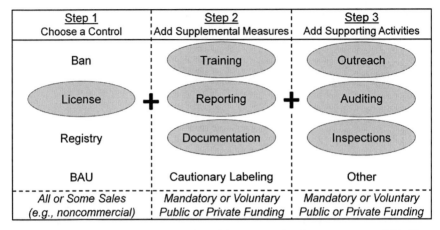

FIGURE 5-1 A schematic representation of the strategy development process. This illustrative example results in a package of policy mechanisms that consists of licensing with training, reporting, documentation, outreach, auditing, and inspections.

would be expected to meet a standard of performance rather than a standard of design or behavior. Performance standards express requirements in terms of outcomes, not the means to achieving outcomes, and give regulated parties the flexibility to achieve regulatory objectives as cost-effectively as possible.[231] For example, a retailer might be expected to train its employees to know when a product is subject to a control; when to ask for and how to check a purchaser's credentials; and how to identify and report suspicious behavior, fraud, theft, and loss; but the retailer would be given the flexibility to conduct the training by whatever means it deemed best.

To judge compliance with a measure or activity, the policy could specify certification and testing requirements, rely on audits or inspections, or employ other tools, potentially left to the judgment of the retailer. In the case of a mandatory provision, the retailer might be held to the standard by a government agency—federal, state, or local; in the case of a voluntary provision, a trade association might provide oversight.

Though businesses might choose to develop their own training materials, a government agency or trade association might provide retailers with training materials, such as pamphlets and videos (see Appendix F), but give retailers leeway to adapt and apply the materials as they see fit. In the discussion of tradeoffs among strategies, later in this chapter, the committee considers how a business's incentives might differ if a measure is voluntary or mandatory.

Just as the committee did not assume a particular assignment of responsibilities for outreach, training, reporting, documentation, or audits and inspections, neither did it assume a particular funding source—private or public—for those activities. In the case of a publicly funded program, either mandatory or voluntary, the government would bear the initial cost, which would fall eventually on taxpayers; in the case of a privately funded program, be it mandatory or voluntary, industry would bear the initial cost, which would fall eventually on some combination of proprietors or shareholders, employees, and customers. In either case, the cost of implementing the measure or activity, if not administering it, should be about the same, if retailers are held accountable to the same standard of performance.

Ultimately, a control strategy would require the specification of threshold concentrations and quantities for each precursor chemical under consideration, but for purposes of this analytical exercise, the committee took the existence of specifications as given, but as yet unestablished. Moreover, the committee did not assess different approaches—bans, licensing, registries, or BAU+—for different precursor chemicals separately; rather, it considered each approach for any or all precursor chemicals. In an actual policy-making process, not all precursor chemicals would need to be treated identically; that is, some Group A chemicals could be subject to a ban, others to a licensing regime, others to registry requirements, and still others to additional mandatory or voluntary measures and activities without any new control. However, the United Kingdom ruled out mixed controls, based on feedback from industry, which preferred the simplicity of a stand-alone licensing regime.

The committee considered the potential for direct benefits, direct costs, ancillary benefits, and unintended consequences, the last of which might reduce benefits or add to costs, as factors to compare across strategies. An ancillary benefit is a favorable impact that is typically unrelated or secondary to the stated purpose of the control, measure, or activity. For example, a control might increase the visibility of transactions and, thus, present information that can be used to develop and improve policy. An unintended consequence is an undesirable security, economic, or social side effect of a policy action. For the most part, these definitions are consistent with those found in OMB guidance, but the committee diverges from OMB by referring to a subset of unintended consequences that lack sufficient basis for ranking as *uncertainties*.[231] Uncertainty, as a term of art among economists, differs from risk, a term that OMB applies, in that it can be used to describe conditions when the future is unknown—in the extreme, anything or nearly anything can happen—and it is impossible to characterize the probability, if not the severity, of a consequence. Thus, an uncertainty is akin to an unknown risk.[232-234] The committee categorized anticipated effects, whether direct, ancillary, or unintended, as either benefits or costs, but treated particular uncertainties, including displacement, separately.

As noted previously, the committee also diverges from OMB guidance by framing most of the benefits of the control strategies as pathways to attaining policy objectives—or capabilities—and not as actual policy outcomes. In those instances, the committee had enough information about the structure of the pathway to support a qualitative, operational comparison across strategies, even if it did not have enough information to characterize the ultimate effect of the strategy. For example, the committee might feel comfortable comparing the stringency of a ban, licensing, or a registry, in terms of its operating principles, but would not feel comfortable comparing the effects of the controls on explosions, deaths, or injuries.

ASSESSING TRADEOFFS AMONG CONTROL STRATEGIES

In this section, the committee presents the results of the analytical exercise, which illustrates one possible approach to examining, qualitatively, the benefits, costs, and uncertainties of strategies and the tradeoffs among them. The committee does not intend to suggest that this is the only possible approach to assessing tradeoffs or that these are the only possible strategies. However, the strategies under consideration encompass many features of control strategies that have been employed in other contexts (see Chapters 3 and 4) and could be employed in this context. Different specifications are also possible; for example, licensing or registry requirements could cover all purchasers, but set different quantity limitations for commercial and noncommercial purchases.

Similarly, the committee makes no claim to have identified all the benefits, costs, and uncertainties that could emerge from a strategy; rather, it addresses the particular types of benefits, costs, and uncertainties that came to the fore in its deliberations.

The committee assessed the strategies slightly differently, depending on whether they included a new control. The committee treated the strategies with new controls—ban, licensing, or registry—as complete packages and assessed them in terms of how they might create a pathway or elicit an outcome in relation both to current policy and to each other. The committee ranked those three strategies with respect to the benefits and costs of each, but, as noted above, it did not attempt to rank them on the basis of the uncertainties (see Appendix H).

The committee treated the BAU+ strategy less as a package and more as a compendium of measures and activities, not strictly comparable to the other three strategies. These measures could be implemented together or separately, reflecting the flexibility of the premise that policy makers could choose to forgo new controls, but still require or encourage additional outreach, training, reporting, or other efforts. Thus, the committee considered individually, and without rankings, the benefits, costs, and uncertainties of several measures and activities that could constitute elements of a BAU+ strategy.

All the assessments, including the benefit and cost rankings of the strategies with new controls, are strictly qualitative. On that basis, the committee cannot say how much better or worse one strategy might be relative to any other, only that it might be qualitatively better or worse with respect to each individual attribute.

The committee did not attempt to convert the ordinal rankings to scores because it would have needed to apply weights to each benefit and cost. It is unlikely that society would value each benefit or cost equally; moreover, the relative importance of different costs, benefits, or uncertainties could change over time.

By design, any benefits, costs, or uncertainties tied solely to a supplemental measure or activity, irrespective of the particular control, would be the same across the strategies that contain new controls, because the measures and activities are fixed across strategies. For example, the ban, licensing, and registry strategies rank equally in terms of awareness because that benefit flows largely from outreach and training, which is the same in each case.

Assessment of Benefits

The committee considered anticipated benefits relating to the capability to impede, deter, or reduce acquisitions of precursor chemicals for illegitimate purposes; the capability to increase awareness of chemicals, concerns, and requirements; the capability to track and correlate suspicious behavior; the capability to provide feedback on implementation; early warnings from vetting; better visibility of transactions and additional information for developing and improving policy; and noneconomic social benefits, some of which could be the same across strategies. These benefits, which include both direct and ancillary benefits, framed largely as pathways to attaining policy objectives or other outcomes, are described in Box 5-1.

Table 5-1 presents a summary of the assessment of benefits, exclusive of any offsets that might emerge from the assessment of particular uncertainties later in this chapter.

The committee ranked the ban, licensing, and registry strategies with respect to the benefits outlined in Box 5-1. The rankings depend on the type of access to precursor chemicals granted under each strategy. For a ban, only commercial purchasers would have access (denoted as C in Table 5-1). The ban would allow commercial purchases with evidence of the purchaser's commercial status and disallow noncommercial purchases. In the case of a license, both commercial purchasers with evidence of commercial status and noncommercial purchasers holding a license would have access (denoted as $C+L$ in Table 5-1). Finally, under the registry, commercial purchasers with evidence of commercial status and noncommercial purchasers who sign a registry and present a government-issued ID would have access (denoted as $C+R$ in Table 5-1). Each strategy also includes outreach, training, reporting, documentation, and audits or inspections (Appendix H).

All three control strategies differentiate between commercial and noncommercial sales and treat commercial sales the same. Thus, the rankings apply only to the merits of each strategy in relation to noncommercial sales, with all else remaining the same. In reality, separating the two could require additional

> **BOX 5-1**
> **Benefits**
>
> The anticipated direct and ancillary benefits of each of the three strategies that contain new controls (a ban, licensing, or a registry) can include, but are not limited to, the following:
>
> - A capability to impede or deter and reduce acquisitions of precursor chemicals through legitimate channels. The ban, as specified, would prohibit all noncommercial purchases, but licensing or a registry, as specified, would allow some noncommercial purchases. The first control (the ban) might impede all noncommercial purchases; the second control (licensing) might impede illegitimate noncommercial purchases; and the third and least restrictive control (the registry) might deter and reduce illegitimate noncommercial purchases.
> - A capability to increase awareness of chemicals, concerns, and implementation mechanisms and requirements, stemming largely from outreach and training.
> - A capability to track and correlate suspicious activity, fraud, theft, and losses, ex ante, and/or investigate incidents, ex post, through data collection and analysis.
> - A capability to provide feedback on implementation to retailers and others from audits and inspections, including mystery shopping. Information might be used for educational purposes with or without an enforcement objective.
> - Better visibility of retail-level transactions and additional information about transactions, largely from reporting, record keeping, and data collection and analysis, for developing and improving policy. The literature on supply chain security (Chapter 3) indicates that visibility and transparency are essential to improving and maintaining supply chain security.
> - Noneconomic social benefits, for example, from enhanced perceptions of safety.
>
> Moreover, in the case of a strategy with licensing, vetting for a license to purchase precursor chemicals could also draw early attention to malicious actors.

credentialing for commercial entities, depending on the extent of ordinary business requirements for a commercial enterprise (operating licenses, permits, etc.), to prevent malicious actors from posing as commercial entities, as occurred in the Oslo bombing. As evidenced by the difficulties the EU has encountered (see Box 4-1), the task of separating commercial and noncommercial users can present challenges.

The strategy featuring the ban (C) might be viewed as the most restrictive of the three strategies; the licensing strategy ($C+L$) is in the middle; and the registry

TABLE 5-1 Strategies Ranked by Benefit Type, Excluding Uncertainties

	Capability to Impede, Deter, or Reduce Acquisitions	Capability to Increase Awareness of Chemicals, Concerns, and Requirements	Capability to Track and Correlate	Capability to Provide Feedback on Implementation	Early Warning from Vetting	Better Visibility of Transactions for Policy	Noneconomic Social Benefits
Most Beneficial	C	C, C+L, C+R	C+L	C, C+L, C+R	C+L	C+L	C
Intermediate	C+L		C+R			C+R	C+L
Least Beneficial	C+R		C			C	C+R

NOTE: C, C+L, and C+R correspond to three different strategies, featuring a ban, licensing, and a registry, respectively, on noncommercial purchases. For additional details, see text and Table H-1 (Appendix H).

strategy ($C+R$) is the least restrictive. Thus, the strategy with the ban (C) would rank highest for capability to impede, deter, or reduce acquisitions and for non-economic social benefits, if stringency conveys perceptions of security, and the strategy with the registry ($C+R$) would rank lowest.

Looking only at the benefits shown in Table 5-1, the licensing strategy ($C+L$) dominates the registry strategy ($C+R$), but the relative position of the ban strategy (C) is ambiguous; in some instances, the ban strategy ranks above the licensing or registry strategies and, in others, it ranks below them. The differences hinge largely on the relative stringency of the strategies, as noted above, on non-economic social benefits, and on flows of information.

The strategy featuring the ban (C) ranks last in terms of both capability to track and correlate and better visibility of transactions. For these benefits, the results depend on the quantity and quality of information that a control might yield, which, in turn, depends on the number of transactions associated with each strategy and the type of information collected in each transaction. Because the ban is the most restrictive control, with the fewest potential transactions, it would necessarily generate the smallest number of observations. Moreover, whatever kinds of commercial information the policy community would gain from the ban, it would also gain from licensing or a registry, because they allow the same commercial purchases as the ban, under the same conditions. In summary, the strategy with the ban would rank last for these benefits because it would provide only a subset of the information that one would expect from licensing or a registry.

Whether a strategy that features licensing would rank above or below a strategy with a registry on either tracking and correlation or visibility of transactions requires further consideration.

Regarding tracking and correlation, one might expect to see the *most* transactions under the least stringent form of control (the registry), but the registry might not convey the *best* information. Licensing is almost certain to provide better information about legal transactions than a registry and, in these contexts, the quality of information matters. Licensing requires vetting beyond that required to obtain a typical government-issued ID and thus will provide more insight to the background of the purchaser and, possibly, the nature of the transaction. The committee chose to rank the licensing strategy ahead of the registry strategy because, in its view, a smaller number of observations with additional relevant details could be more useful analytically than a larger number of observations that lack those details.

With regard to transaction visibility, the difference between licensing and a registry might depend partly on whether licensees must provide a justification for their license applications. If a licensee must provide a statement of need (e.g., use in a particular hobby) to obtain a license and if the transaction record includes the license number, then each transaction can be tracked to a need. However, the committee warns against overreaching for data. Relevant details come at a price to both the provider and the collector, and the more details that a program demands from participants, the costlier it might become to implement and administer.

The benefits relating to awareness and feedback are the same across strategies, because each strategy includes the same provisions for outreach, training, and audits and inspections; however, the benefits of vetting apply only to licensing, because it is the only control that requires vetting.

Assessment of Costs

Table 5-2 presents a summary of the assessment of costs, exclusive of any losses that might emerge from the assessment of particular uncertainties later in this chapter. The first three costs are direct costs, whereas the last four are rankable, unintended consequences. Looking only at the costs shown in Table 5-2, none of the strategies clearly dominates any other.

The committee considered anticipated costs relating to public-sector expenditures on administration and implementation; public-sector expenditures on law enforcement; private-sector expenditures on administration and implementation; forgone sales and surplus; forgone use and surplus; additional transaction time; and noneconomic social costs. These costs, which include both direct costs and unintended consequences, are presented in Box 5-2.

The public- and private-sector costs of administering, implementing, and enforcing each strategy should be about the same in terms of the commercial component (i.e., the requirement for evidence of commercial status). Thus, any differences in these costs, by strategy, would reside in the additional requirements that would accompany licensing or a registry.

The strategy featuring the ban on noncommercial purchases (*C*) looks best in terms of expenditures on administration, implementation, and law enforcement, but shows relatively poorly in terms of forgone sales, forgone use, and noneconomic social costs. The ban has the fewest program features and, thus, should be the least costly to carry out; however, because of its relative stringency, it is expected to have the most noticeable effect on sales. If (1) all three strategies result in the same number of commercial transactions because the requirements for commercial sales are the same, (2) the ban results in the fewest noncommercial transactions, and (3) the registry results in the most, then it seems plausible that the ban would entail the largest losses of producer and consumer surplus and the registry would entail the smallest. Similarly, the public might feel most constrained under a ban and least constrained under a registry requirement, resulting in noneconomic social costs related to the losses of personal or societal freedom.[235]

A strategy with licensing might incur the highest administration and implementation costs because issuing licenses (application and approval) and collecting and analyzing license-related data would absorb additional resources. However, the distribution of those costs to the public or private sector would differ, depend-

TABLE 5-2 Strategies Ranked by Cost Type, Excluding Uncertainties

	Administration and Implementation (Public)	Law Enforcement	Administration and Implementation (Private)	Forgone Sales and Surplus	Forgone Use and Surplus	Additional Transaction Time	Noneconomic Social Costs
Most Costly	C+L	C+R	C+L	C	C	C+R	C
Intermediate	C+R	C+L	C+R	C+L	C+L	C+L	C+L
Least Costly	C	C	C	C+R	C+R	C	C+R

NOTE: C, $C+L$, and $C+R$ correspond to three different strategies, featuring a ban, licensing, and a registry, respectively, on noncommercial purchases. For additional details, see text and Table H-1 (Appendix H).

BOX 5-2
Costs

The anticipated direct costs and unintended consequences of each of the three strategies that contain new controls (a ban, licensing, or a registry) can include, but are not limited to, the following:

- Public-sector expenditures on licenses, administration, outreach, training, data intake and analysis, and audits and inspections. Expenditures here and in the following bullets include time and materials and capital outlays, for example, for record-keeping systems. If licenses for noncommercial purchases are fee based, then public-sector expenditures on licensing and related administration would be limited to those exceeding the fee.
- Public-sector expenditures on law enforcement. If retailers report fraud (i.e., fraudulent evidence of commercial status, license, or ID), suspicious behavior, or additional theft and loss, then law enforcement agencies would incur additional costs.
- Private-sector (business and consumer) expenditures on licenses, administration, outreach, training, reporting, documentation, and audits and inspections.
- Forgone sales and associated losses of economic surplus. Surplus among sellers amounts to the difference between the price a seller would be willing to accept for a product and the price a seller actually receives. It is roughly akin to profit. If sales decline because of a control, that surplus might also decline. Costs to businesses might be passed on to proprietors, shareholders, employees, and consumers.
- Forgone use and associated losses of economic surplus. Surplus among buyers amounts to the difference between the price a buyer would be willing to pay for a product and the price a buyer actually pays. A buyer might have been willing to pay $2.50 for a bottle of cleaning product, but the market price might only be $2.25. If a control limits purchases, buyers will tend to accrue less economic surplus.
- Additional transaction time at the point of purchase, stemming partly from requirements to present evidence of commercial status, a license, or a government-issued ID.
- Noneconomic social costs, for example, from perceived losses of personal or societal freedom.

If the costs of forgone sales and use were framed as opportunity costs, they would amount to the difference in the value of alternatives to sellers and buyers along the supply chain.

ing on whether licensing fees covered some of those activities. Licensing, unlike a ban or registry, might also entail outreach and training on the license.

In the case of law enforcement, the rankings of the strategies depend largely on differences in opportunities for fraud, resulting from differences in credential requirements and numbers of transactions. (This assessment assumes that the potential for theft and loss are about the same across options and that a federal hotline continues to handle some or most concerns about suspicious behavior, but not necessarily all.) Strategies that feature licensing or a registry are likely to present more opportunities for fraud than a strategy with a ban because they require a credential. Lacking evidence on whether fraud would be more likely with licensing or a registry, the committee assumed that attempts to deceive—and detection—would be equally likely in either case. Thus, the ranking $(C+R > C+L)$ assumes that a registry elicits more transactions than licensing and, as a consequence, generates a larger number of calls to police.

Similarly, the ranking for transaction time $(C+R > C+L)$ depends primarily on the relative number of transactions, but the registry might also be costlier because it requires not only the presentation of a credential, but also a signature.

Consideration of Uncertainties

The committee explored separately the potential for noncompliance among retailers, institutional amnesia and employee turnover, outright circumvention, unintended knowledge transfer, displacement, over-implementation, commercial disruption, and discriminatory profiling as relative unknowns that could affect the benefits and costs of controls. Had these consequences been less uncertain, they would have been incorporated into the foregoing benefit and cost assessments as either offsets to benefits or additional costs. Specifically, noncompliance, institutional amnesia and employee turnover, outright circumvention, unintended knowledge transfer, and displacement would have been handled as offsets to benefits; over-implementation, commercial disruption, and discriminatory profiling might be handled as additional costs, though OMB suggests treating all unintended consequences as benefit offsets. These uncertainties are described in Box 5-3.

The committee could not rank the control strategies for each type of uncertainty because of the complexity of the task and a lack of evidence. Nevertheless, the committee offers these observations on the potential tradeoffs among the strategies in this domain.

First, it seems plausible that concerns about circumvention, on the one hand, and displacement and commercial disruption, on the other, would be inverted; in particular, circumvention might be of least concern for a strategy with a ban and of most concern for a strategy with a registry, while concerns about displacement and commercial disruption might be highest for a ban and lowest for a registry, at least insomuch as probability is a focal point of concern (see equations in Appendix B). As a matter of logic, the control that is most stringent and least

> **BOX 5-3**
> **Uncertainties**
>
> The controls under consideration present uncertainties that can offset benefits or add to costs under a variety of circumstances, including the following:
>
> - Retailers might not comply, either inadvertently or intentionally, because compliance is too difficult or costs too much relative to the threat of penalties or concerns about market response, culpability, and liability, each of which could have substantial financial implications.
> - Institutional amnesia and employee turnover could undermine implementation if a company forgets its obligations as time passes, or cannot keep pace with a rapidly changing roster of employees, as might be typical of some retail environments. Institutional amnesia and employee turnover could be treated as a subset of noncompliance.
> - Outright circumvention, by comparison, would occur, for example, if a terrorist intentionally bypasses a control to obtain a precursor chemical without authorization through diversion from legitimate channels or document falsification.
> - Unintentional knowledge transfer would occur if adopting a control alerts terrorists to the potential utility of a chemical as a precursor or confirms suspicions of utility.
> - Displacement refers to the possibility that terrorists will adopt new modes of attack if obtaining precursor chemicals becomes too costly or difficult. Displacement would also include a shift among precursor chemicals; that is, a terrorist might obtain a different chemical to produce an IED, because his or her first choice becomes less readily available.
> - Over-implementation, a term of the committee's devising, would occur, for example, if retailers request unnecessary credentials, such as a license or government-issued ID from a commercial buyer; turn back legitimate credentials; or apply controls to unrestricted products.
> - Commercial disruption might occur if a control were so burdensome as to drive businesses—producers, distributors, or retailers—out of the market or to deny supplies of a precursor chemical to an industry with immediate and legitimate need. If a control were sufficiently burdensome (e.g., in terms of complexity or cost) or resulted in too few sales, it could lead businesses to cease production, distribution, or sale of a particular chemical or, in the extreme, to fold.
> - Discriminatory profiling would occur if retailers base assessments of suspicious behavior on inappropriate visual or other cues, such as complexion, attire, or accent, that might relate to gender, race, ethnicity, or religion. Profiling can, itself, create vulnerabilities if terrorists anticipate profiling and work around it, for example, by recruiting individuals who will not raise suspicion.

susceptible to circumvention (specifically, the ban on noncommercial sales) might also be most likely to engender displacement and disrupt commerce, by giving terrorists a reason to try working with other chemicals or pursue other modes of attack, and by preventing legitimate sales and use, respectively.

Second, it is possible that concerns of retailer noncompliance, distinct from outright circumvention, would increase with complexity and administrative costliness, in which case the strategy that is easiest to implement might be the least likely to result in noncompliance. On that basis, a ban, which might be the least cumbersome of the three controls, might result in less retailer noncompliance than either licensing or a registry; however, the ranking would depend ultimately on the point of failure and its cause, for example, transaction time, implementation costs, or other factors.

Third, concerns about over-implementation might differ for commercial and noncommercial purchasers. For a commercial purchaser, a ban on noncommercial sales might be the most worrisome control and the registry the least, not because bans are more likely to result in more over-implementation than registries, but because the consequences could be more serious. In the case of a registry, a retailer might needlessly request a signature and ID, but the request would not necessarily block the sale; however, in the case of a ban the retailer might flatly—and inappropriately—deny the sale. Under the ban, noncommercial purchasers do not face concerns about over-implementation, because they are not eligible to purchase precursor chemicals in the first place, but both the licensing and registry requirement could result in excessive denials, for example, if retailers reject valid licenses or IDs. There might be more purchases under a registry than with licenses and, hence, more opportunity for denials, but whether retailers are more or less likely to reject a license or a government-issued ID is an open question; thus, the committee has no basis for distinguishing between the two controls.

Fourth, any of the strategies could lead to discriminatory profiling, because retailers have a right to refuse sales, whether they are noncommercial or commercial, if they encounter suspicious behavior. A retailer might, however, use the authority to turn away a legitimate buyer based on the retailer's own biases. Two potential differences are that a strategy that features a registry might entail (1) more transactions and, thus, present more opportunities for profiling than a strategy that features either a ban or licensing and (2) a softer requirement for noncommercial purchasers and, thus, leave more room for subjectivity, which might also allow more profiling.

Fifth, the committee found no basis for distinguishing among the strategies with regard to concerns about institutional amnesia, employee turnover, or inadvertent knowledge transfer.

Assessments of Other Measures and Activities

The committee also considered several measures and activities that could accompany existing controls as elements of a BAU+ strategy and operate under legal mandates or through voluntary participation, with varying degrees of government and industry involvement and oversight. Specifically, it looked separately at opportunities for outreach; training and reporting on suspicious behavior, theft, and loss; documenting transactions with electronic record keeping; and auditing with mystery shopping, partnered with training, reporting, or other measures. As noted previously, the committee also considered cautionary labeling, but did not include the measure in any of the strategies, including BAU+, because of concerns about unintentional knowledge transfer.

The results of the committee's assessments, which are presented under BAU+ in Table H-1, suggest that some gains in security and improvements in flows of information are possible without a ban, licensing, or a registry, but that BAU+ would still entail costs and uncertainties. For example, training and reporting on suspicious behavior, fraud, theft, and loss could create better awareness of chemicals, concerns, and implementation mechanisms and requirements; a capability to deter and reduce illegitimate acquisitions; a capability to track and correlate suspicious activity and investigate incidents; better visibility of retail-level transactions for developing and implementing policy; and some noneconomic social benefits. However, it would also entail public- and private-sector costs of administration and implementation, including data intake and analysis, and, to a lesser extent, many of the same kinds of uncertainties as the three strategies with new controls.

Summary of Assessments and Tradeoffs

At the start of this chapter, the committee indicated that it would consider tradeoffs among control strategies in relation to three policy objectives, namely restricting malicious actors' access to precursor chemicals, gathering information that could be used to prevent future attacks or investigate prior attacks, and minimizing burdens on industry, commerce, and legitimate users. The foregoing assessments illustrate just one possible approach to comparing strategies; nevertheless, they shed light on potential economic and noneconomic tradeoffs and the conditions that might affect them. The committee was able to rank the strategies in terms of individual benefit and cost attributes, but it cannot say how much more or less beneficial or costly one strategy is relative to the next; moreover, the final standing of the strategies would depend partly on whether and how uncertainties enter the assessment.

The tradeoffs among benefits, without yet considering the cost categories or uncertainties, are reasonably clear (Table 5-3). The strategy that features a registry never surpasses the strategy that features licensing for any of the anticipated benefits, thus leaving policy makers to consider the relative merits of licensing

TABLE 5-3 Control Strategy Rankings by Benefit

Rank of Strategy by Benefit		Ban	License	Registry
	1st	• Capability to impede, deter, or reduce acquisitions • Noneconomic social benefits	• Capability to track and correlate • Visibility over transactions to inform policy • Early warning from vetting	
	2nd		• Capability to impede, deter, or reduce acquisitions • Noneconomic social benefits	• Capability to track and correlate • Visibility over transactions to inform policy
	3rd	• Capability to track and correlate • Visibility over transactions to inform policy		• Capability to impede, deter, or reduce acquisitions • Noneconomic social benefits

NOTE: Summary of rankings of strategies with new controls by benefit type (not including costs and uncertainties). First is best or most beneficial, and third is worst or least beneficial. The order of the benefits within each rank (first, second, or third) has no analytical significance.

and a ban. Generally speaking, the ban would block more sales and, thus, looks better as an impediment to access, whereas licensing would allow more sales and, thus, looks better as a source of information, especially with vetting for licenses. However, upon introducing the committee's observations on the uncertainties, the tradeoffs in benefits become murkier. A strategy that features a ban might be less likely to result in noncompliance or circumvention than one that features licensing, but it might be more likely to yield displacement. Terrorists who cannot obtain precursor chemicals could opt for more or less damaging modes of attack; if the former, the control could make matters worse. Even if the new mode is no worse than the old, it is possible (as might be said of the pseudoephedrine example described in Box 3-2) to solve the immediate problem, in this case, access, without solving the ultimate problem, terrorism, and to incur considerable cost in the process.

The comparison of costs is less clear (see Table 5-4). The strategy that features the ban ranks best on expenditures for implementation, administration, and enforcement and on transaction time, but ranks worst, because of its stringency, on forgone sales, forgone use, and noneconomic social costs. In addition, the potential for commercial disruption—which bears directly on sales and use—might increase with stringency, which would reinforce the negative results, whereas the potential for over-implementation could push in either direction, depending on commercial status. Licensing occupies the middle ground on enforcement expenditures, forgone sales, forgone use, noneconomic social costs, and transaction time, but ranks poorly on implementation and administration expenditures. The strategy that features the registry ranks best on forgone sales, forgone use, and noneconomic social costs; it occupies the middle ground on expenditures for implementation and administration; and it ranks worst for

TABLE 5-4 Control Strategy Rankings by Cost

Rank of Strategy by Cost		Ban	License	Registry
	1st	• Administration/implementation expenditures (private and public) • Law enforcement expenditures • Additional transaction time		• Forgone sales and use • Noneconomic social costs
	2nd		• Law enforcement expenditures • Forgone sales and use • Noneconomic social costs • Additional transaction time	• Administration/implementation expenditures (private and public)
	3rd	• Forgone sales and use • Noneconomic social costs	• Administration/implementation expenditures (private and public)	• Law enforcement expenditures • Additional transaction time

NOTE: Summary of rankings of strategies with new controls by cost type (not including benefits and uncertainties). First is the least costly, and third is the most costly. The order of the costs within each rank (first, second, or third) has no analytical significance.

transaction time and enforcement expenditures. Cost by cost, the strategy that features the registry ranks somewhat better than the strategy that features licensing in most, but not all, regards.

Among the three control strategies, the assessment suggests a tradeoff between cost types: The strategies that rank well on the direct costs (expenditures on implementation, administration, and enforcement) rank less well on the unintended consequences (forgone sales, forgone use, noneconomic social costs, and transaction time), and vice versa.

Overall, the assessment of benefits, costs, and uncertainties suggests that the benefits of stringency might come at the price of lost economic surplus to both sellers and buyers, noneconomic social costs, displacement, and commercial disruption. As OMB instructs (see Appendix G), a good regulatory analysis must consider ancillary benefits, unintended consequences, and what the committee refers to as uncertainties.[231] To choose a strategy, the policy community would need to flesh out the details of each strategy and, to the extent possible, put numbers into the analysis.

Finally, whether a measure or activity is mandatory or voluntary might affect the likelihood of a retailer's compliance and, hence, the efficacy of any strategy. If a measure or activity is mandatory—required under federal, state, or local law or ordinance—compliance incentives could flow from threats of fines or imprisonment as well as concerns about market responses, moral culpability, and liability. By contrast, a voluntary program would not a convey a statutory threat, but concerns about market responses, moral culpability, and liability would persist. A retailer might not want to be seen as a business that enables terrorism or to learn that it has enabled terrorism, if it might lose sales from reputational damage, do harm to others, or bear financial responsibility.

In addition, an active trade association—for example, one that offers outreach, training, reporting, or other programs—can introduce reinforcing incentives to a voluntary program. For example, trade associations can and have conditioned membership and benefits on compliance, highlighted members' accomplishments, issued public reports on members' performance, and adopted other tactics that shine light on members' behavior (Chapter 3). Still, a mandatory measure or activity, under force of law, is likely to carry more weight than a voluntary measure or activity.

CONCLUSION

Thus far, this chapter has set aside some of the practical concerns raised in other chapters that could be relevant to the assessments at hand. In Chapter 4, the committee presented evidence from the European Union (EU),[191] United Kingdom, and elsewhere on implementation challenges that relate to market breadth and reach, customer identification, and harmonization. Although the EU has not yet accumulated enough experience to allow the committee to speak definitively to these challenges, it has accumulated enough experience to suggest implications for U.S. policy makers.

The EU and EU member states (MS) have had difficulty reaching all retail outlets because precursor chemicals are constituents of wide-ranging, broadly distributed products that are sold at brick-and-mortar outlets and online. Although European officials have been working with trade associations and engaging with internet retailers to address potential vulnerabilities, it might not be possible to reach everyone or address all the gaps fully. The United States presents a similar retail landscape, so similar problems can be expected. Moreover, as noted in Chapter 4, European officials have faced difficulty distinguishing between commercial and noncommercial purchasers and also between eligibility and need. If U.S. policy makers were to attempt to distinguish between commercial and noncommercial purchasers, they would need to carefully consider means of implementation.

Harmonization is, perhaps, more complicated. The European legislation, albeit applicable to all EU MS and thus "harmonized" as an overarching regulatory framework, provides the states with considerable flexibility by leaving open decisions about how to restrict the general public's access to certain precursor chemicals, be it through a ban, licensing, or a registry. At present, most of the 28 EU MS have complied with the legislation, but they have done so differently and without reciprocity, which has presented challenges for business seeking to engage in interstate commerce and might create opportunities for terrorists to game regulations across EU MS.

Policy makers in the United States might face similar concerns about jurisdictional difference, both in contemplating federal, state, or local controls and in developing or encouraging related mandatory or voluntary measures and activities. For example, several states already restrict access to ammonium nitrate (AN), but they do so by different means. The committee did not conduct a thorough review of legislation across states, but did observe a lack of uniformity in state law. Evidence obtained from a regulatory body in one state suggests that differences in states' rules might lead to confusion among AN distributors, retailers, and purchasers that could interfere with commerce—across or within states—and undermine one or another state's efforts to mitigate risks. The evidence, both foreign and domestic, suggests a potential role for national coordination, if not federal legislation, as a matter of commerce and security.

The committee offers three final comments on policy design and analysis: more specifically, on providing flexibility to leverage experience, on harvesting lessons-learned to leverage experience, and on the need for future analytical efforts.

First, experience that is fed back into policy design and implementation can be used to improve both of these dimensions of policy, but only if policy makers can revisit or even re-craft some aspects of whatever control strategies they choose initially. They cannot leverage the experience if the strategies are unalterable. Each of the foregoing control strategies is defined generally and can be amended insomuch as policy makers can add or subtract precursor chemicals, as they have in Europe (Chapter 4); can curb or extend provisions to cover a smaller or larger slice of society, as the EU is contemplating; and can add or subtract supplemental measures and activities. The previous Academies' study suggested shifting from one level of stringency to another with changes in the policy environment, but wholesale adjustments might be difficult, in practice.[14] It could, however, be easier to shift from a less stringent set of requirements to a more stringent set in response to a change in circumstances.

Second, as a related matter, policy makers cannot leverage experience if they cannot harvest it through evaluation or other means. Developing metrics, measures, or other tools to assess the effectiveness of a control strategy that is already in place presents methodological and practical challenges much like those of retrospective regulatory assessment (Appendix G). For example, it might be difficult to separate the effects of a control strategy from the effects of everything else that is happening in a particular environment, in part because little is known about the underlying risks and data are scarce or sensitive. A terrorist event might not occur because a control strategy worked, because such events rarely occur in any case, or because the terrorists chose a different course of action.

Notwithstanding the analytical challenges, the policy community could take a step toward evaluation by reviewing program implementation, treating the EU's experience with its control strategy as a test case, and, potentially, experimenting with a broader range of methodologies. Program participation rates and results from audits and inspections, including mystery shopping, can be used, for example, to review the extent to which controls, measures, and activities are up-and-running and how they are playing out, at least administratively. Moreover, the EU's continuing experience with bans, licensing, and registration might provide fertile ground for an analysis of those types of controls. As time passes, it might be possible to learn more about what is or is not working in the EU, the related costs and unintended consequences, and the applicability of lessons learned to U.S. policy making. Obviously, the EU and United States face substantially different threats and differ in many social, political, and economic regards, but it might still be possible to learn from the EU's experience.

In addition, it might be possible to make better—or additional—use of data on terrorist episodes, not only for purposes of regulatory impact assessments, but also for conducting other types of supporting analyses.[236,237] Approaches might include, for example, applications of fault trees and dynamic behavioral models that can provide insight to terrorists' and others' responses to policy and threats. Even if some data must be held closely because of their sensitivity, data still can be used in rigorous analyses that are subject to peer review, criticism, and scrutiny, in accordance with appropriate restrictions on handling and distribution. Work to address the limitations of current analytical methods (Appendix G) could, eventually, help to support a more complete assessment of possible controls.

Third, this report constitutes a starting point, not an ending point, for analyzing possible control strategies. Whereas this report sets out an analytical framework and draws insight from a notional, qualitative assessment of benefits, costs, and uncertainties, a full consideration of specific regulatory or other actions would require more time, data, industry participation, and specificity, including a clear articulation of the structure and content of proposed regulatory or other actions. With more *deliberative thinking* now, the policy community might avoid the pitfalls of *intuitive thinking* later.

6

Potential Approaches to Restricting Malicious Actors' Access to Precursor Chemicals: Conclusions and Recommendations

In this report, the committee has presented its assessment of the threats posed by precursor chemicals used to make homemade explosives (HMEs) that can be employed in impovised explosive devices (IEDs), the challenges of restricting malicious actors' access to precursor chemicals as they move through the supply chain from manufacturers to end users, and the relative merits of different approaches to addressing vulnerabilities without undermining legitimate commerce and use. Responsive to the statement of task, this chapter provides recommendations that draw from the findings and conclusions of previous chapters.

Precursor chemicals have played a significant role in the manufacture of HMEs used in IED terrorist attacks (see Chapter 1), but that role has changed over time and will continue to change in the future. Malicious actors intent on damaging critical infrastructure, inflicting causalities, and eliciting fear have demonstrated the ability to change their tactics in response to the implementation of controls or shifts in the availability of particular precursor chemicals (see Chapter 2). Given the ease of accessing precursor chemicals and the information needed to manufacture HMEs, particularly via the internet, the threat of terrorist attacks using IEDs within the United States has not abated.

While the United States has not experienced a major domestic attack with a vehicle-borne IED (VBIED) since the 1990s, the committee stresses the importance of deliberative thinking before crises, and cautions against intuitive thinking and action bias during and after crises. Even if event-driven policy making is unavoidable, it might be possible to lay a foundation for constructive policy responses. The analyses outlined in this report should serve as the basis for such measured actions.

The committee also stresses the importance of periodically reevaluating priorities among precursor chemicals, in light of changes in the threat environment, and of designing control strategies with means of harvesting and leveraging data to learn from experience and strengthen policy over time. The committee assigned chemicals of particular concern to one of three groups, A, B, or C, by priority, but policy makers should treat neither the groups nor any control strategy as static. Policy makers should view policy development and implementation as part of an ongoing process of risk management (see Appendix B) and, for that reason, should consider designing any control strategy with enough flexibility to accommodate environmental change, learning, and amendments.

As described in Chapter 3, the lack of consistency at the federal, state, and local levels appears to have resulted in a domestic policy environment that lacks coherence along the supply chain, leading to confusion for those subject to existing controls and to gaps in visibility and oversight. There have been some efforts to improve coordination and consistency as seen with the Joint Program Office. Further coordination among federal, state, and local agencies, and across independent and overlapping governmental and nongovernmental programs, can strengthen policy implementation.

BEYOND PRECURSOR CHEMICALS

The statement of task for this study clearly focused the efforts of the committee on examining precursor chemicals that can be used to produce HMEs for IED attacks. While precursor chemicals are used to produce HMEs for IEDs, the committee recognizes that, in the United States, a substantial majority of explosives incidents as defined by the U.S. Bomb Data Center (USBDC) used commercial explosives, black powder, smokeless powder, and pyrotechnic filler, perhaps due to their ease of legal acquisition. Thus, any strategy that focuses only on access to precursor chemicals cannot eliminate the threat of IED attacks as long as these other explosive materials remain accessible to malicious actors.

Looking through a wider lens, the committee also recognizes the potential for displacement beyond the realm of explosive materials. Even if policy makers were to eliminate the threat of IED attacks, terrorists could shift strategies and tactics, and choose other lethal means or weapons over explosive materials. Recent attacks targeting pedestrians, shoppers, and others with knives, firearms, and vehicles in the United States, Europe, and elsewhere suggest a range of viable options for inciting terror.

RECOMMENDATIONS

Pursuant to the primary goal of reducing the threat of IED attacks by restricting access to precursor chemicals, and cognizant of the need to preserve legitimate commerce and use, the committee details six recommended courses of action and four research areas meriting future attention.

Priority Precursor Chemicals

Recommendation 1: Federal, state, local, and private sector entities attempting to reduce the threat of IED attacks by restricting access to precursor chemicals should focus on both person-borne and vehicle-borne IEDs.

Precursor chemicals can be and have been used in the explosive charges applied in VBIEDs, person-borne IEDs (PBIEDs), aircraft bombings, and detonators. Prior research efforts, such as the 1998 Academies report,[14] and much of current U.S. policy (see Chapters 1 and 3) have tended to emphasize the threat posed by VBIEDs requiring the use of large charges over the threat posed by PBIEDs requiring the use of smaller main charges. However, as shown in Chapter 2, PBIEDs represent a substantial proportion of IED attacks, both domestically and internationally, and can inflict substantial casualties when deployed near crowds of people. The committee concluded that any effort seeking ultimately to reduce the threat of IED attacks should account for precursor chemicals used in PBIEDs along with those used in VBIEDs. Taking a more comprehensive approach to threat reduction would require consideration of the wider range of precursor chemicals that can be used to construct a PBIED, the overlap of those chemicals with VBIED-relevant precursor chemicals, and the masses and properties of the precursor chemicals required to construct a main charge for each attack mode.

Recommendation 2: Federal, state, local, and private sector entities attempting to reduce the threats from person-borne and vehicle-borne IEDs should consider multi-chemical, rather than single-chemical, strategies.

Data from domestic and international terrorist incidents involving explosives demonstrates that a multitude of precursor chemicals can be used to manufacture HMEs for VBIEDs or PBIEDs. A strategy that focuses on merely one chemical, such as ammonium nitrate (AN) as proposed by the ANSP, cannot adequately reduce the threat of IED attacks due to the availability of alternatives and the historically demonstrated capacity of terrorists to modify their tactics in response to single-chemical controls. This behavioral response has been observed in Northern Ireland, where a strategy that responded narrowly to particular circumstances focused on individual chemicals allowed the Provisional Irish Republican Army (PIRA) the opportunity to change its tactics while maintaining the amount of damage caused (see Chapter 2).

The committee developed a prioritized list of precursor chemicals according to three equally weighted criteria: (1) whether the precursor chemical can be used to make HMEs suitable for a VBIED, a PBIED, or both; (2) the precursor chemical's history of use in manufacturing HMEs; and (3) whether the precursor chemical can be used independently, or is dependent on other specific chemicals, for creating a main charge. On that basis, the committee established three groups of chemicals,

Groups A, B, and C, by order of priority. Chemicals that met all three criteria—that is, they could be used for both VBIEDs and PBIEDs, had a history of use, and could be used independently—were placed into Group A. Chemicals that satisfied two criteria were placed in Group B, with one exception, and chemicals that satisfied one criterion were placed in Group C.

Twenty-eight precursor chemicals are distributed among the three groups. Group A consists of aluminum (powder, paste, and flake), AN, calcium ammonium nitrate (CAN), hydrogen peroxide, nitric acid, nitromethane, potassium chlorate, potassium perchlorate, sodium chlorate, and urea ammonium nitrate (UAN) solution; Group B consists of calcium nitrate, hydrochloric acid, potassium nitrate, potassium permanganate, sodium nitrate, sodium nitrite, sulfur, sulfuric acid, urea, and zinc (powder); and Group C consists of ammonium perchlorate, antimony trisulfide, hexamine, magnalium (powder), magnesium (powder), pentaerythritol, phenol, and potassium nitrite.

The committee made only one exception to the application of the three criteria. While UAN, as a commercially available product, does not meet the historical usage criterion, and thus would otherwise belong in Group B, the committee observed a well-documented history of explosive production in Iraq from analogous solutions of urea and nitrate salts. The history of nitrating urea solutions, the ease of producing various explosives from this precursor, and the strong belief that UAN solution may potentially constitute a significant future threat justified its inclusion in Group A.

The chemicals in Groups B and C could become higher priority chemicals at any time because of their suitability for making HMEs used in IEDs. A change in the threat environment could also cause an increase in amateur production of the precursor chemicals that appear in Groups A, B, and C. As a result, continuing evaluation and, potentially, re-prioritization of chemicals will be required to meet changes in the threat environment.

Strategies at the Retail Level

Recommendation 3: Federal, state, local, and private sector entities attempting to reduce the threats from person-borne and vehicle-borne IEDs should focus on retail-level transactions of precursor chemicals, especially e-commerce.

The precursor chemicals, regardless of their association with Group A, B, or C, flow through complex supply chains and have legitimate uses for both commercial and personal use. Most of these precursor chemicals are readily available through physical retailers, online retailers, or both, and all Group A chemicals are available online. Notwithstanding considerable differences in the paths that each precursor chemical follows, the committee found sufficient commonality to create a generalized supply chain; to overlay existing mandatory and voluntary policy mechanisms that touch the chain; and to identify potential vulnerabilities, as evidenced in a lack of either visibility or oversight.

Federal controls on precursor chemicals focus on their import, manufacturing, and storage and on bulk distribution and transportation, with authorities spread piecemeal across multiple federal agencies. Similarly, trade associations work primarily with bulk quantities of precursor chemicals or just with specific retail groups (e.g., in the agricultural sector).

The committee found a pronounced lack of visibility and oversight in retail-level transactions, especially those involving e-commerce, suggesting ample opportunity for malicious actors to acquire precursor chemicals for making HMEs. Challenges posed by e-commerce, in particular, include the reach of internet sales across national, state, and local borders, the anonymity of the purchaser, the absence of information on the intended use of the chemical, and the lack of opportunity to engage directly with purchasers, as a means of identifying suspicious activity. What could be suspicious behavior, for example, might only manifest through face-to-face engagement. The European Union (EU) and some of its member states (MS) have expressed similar concerns about the challenges of e-commerce. Another challenge is the difficulty of tracking multiple purchases from different sources, both brick and mortar and online. While the committee cannot provide guidance on how to detect the acquisition of combinations of chemicals that can be used in HMEs, the committee acknowledges that having this information could play an important role in the prevention of bombing events.

In addition to concerns about the anonymity of the purchaser using e-commerce, many representatives of industries and trade groups reported that they do not know how precursor chemicals flow to online retailers. For example, none of the AN producers sells its product to exploding target kit manufacturers, yet those finished products are available for purchase, which indicates a gap in the visibility of the overall AN supply chain that the committee could not illuminate. While manufacturers are able to provide some fundamental information on the distributors with which they have direct contact, the practices of smaller distributers or wholesalers are more opaque.

Many options are available for improving visibility and oversight of precursor chemicals at the retail level; some would require changes to law and corresponding regulatory action while others would not. The committee considered opportunities for new mandatory restrictions regarding access to precursor chemicals (i.e., controls) and addressed the possibility of other, potentially supplemental measures and activities—such as outreach, training, and reporting—which can be implemented with or without legal mandates. Further details about other measures and activities are discussed under the rubric of voluntary measures and programs (see Recommendation 6) because these programs have precedent and established records of participation in the private sector, not because they must be voluntary. As stated above, these other measures and activities could be mandatory.

The committee's review of the domestic and international policy landscapes (see Chapters 3 and 4) suggests the possibility of at least four general types of control strategy for retail-level transactions; three of them would add a new mandatory restriction or control and one would not. In particular, the three strategies with new controls would implement either a ban, licensing, or a registry, and the fourth strategy, which the committee refers to as "business as usual plus" (BAU+) would augment existing federal, state, or local controls with other measures and activities. Indeed, any of the four approaches, not just BAU+, could include other measures and activities, established as mandatory or voluntary policy mechanisms, with varying degrees of government and industry involvement.

However, all the strategies would entail costs, including potential adverse effects and or impacts on legitimate commerce and use; benefits, primarily relating to security; and various uncertainties. Overall, the strategies might or might not yield a positive balance. Preliminary evidence suggests that retail-level controls in the EU, which have included bans, licensing, and registries, have decreased the amount of precursor chemicals available on the market and increased the capacity of authorities to investigate suspicious incidents, but that industry and users have experienced some adverse effects. The committee believes that these options would require a thorough evaluation, involving a more rigorous and more detailed analysis (see Recommendation 5) to support any regulatory decisions.

The options for controls are outlined here for the specific case of retail-level transactions and, while potentially providing a means of categorizing options at other levels of the supply chain, may or may not apply to those levels in their current forms. The potential for application elsewhere in the supply chain would require additional and separate consideration given the different nature of the commerce and actors at those locations. In addition, each strategy will require unique caveats and structures to address the challenges posed by e-commerce. In Chapter 5, the committee assessed tradeoffs among control strategies and, for the sake of minimizing disruptions to commerce, focused on applications of any new controls to noncommercial purchases. A summary of the four options is as follows:

- **Banning sales of specific precursor chemicals**. This option could ban the sales of certain precursor chemicals in specific quantities, concentrations, or formulations. The option might also include a right to refuse sales under suspicious circumstances.
- **Requiring a license to purchase**. This option could require a valid license to purchase certain precursor chemicals in specific quantities, concentrations, or formulations. Licensing could involve vetting of the purchaser, such as a background check and verification of need. This option might also include a right to refuse sales under suspicious circumstances. To imple-

ment a licensing scheme for online purchasing, e-retailers could ask purchasers to enter a verifiable license number prior to proceeding to checkout.
- **Maintaining a registry of purchases.** This option could require a signature for purchases of certain precursor chemicals in specific quantities, concentrations, or formulations, and showing proof of identity, such as a government-issued photo ID. This option, like the others, might also include a right to refuse sales under suspicious circumstances. To implement a registry scheme for online purchasing, e-retailers could ask purchasers to enter an electronic signature and a verifiable government-issued ID number prior to proceeding to checkout.
- **Supplementing current controls with other measures and activities.** This option would augment current federal, state, and local controls with various measures and activities, such as outreach, training, and reporting, as discussed below.

In each case, if a control pertains only to noncommercial sales, as in the assessments in Chapter 5, commercial purchasers might be required to provide evidence of commercial status. Separating commercial and noncommercial users could be challenging, as evidenced by the EU's experience, and would require additional analysis to determine whether it would require any new credentialing and to assess the associated benefits, costs, and uncertainties.

Recommendation 4: Federal, state, local, and private-sector entities should explore strategies for harmonizing oversight of the sale and use of commercially available kits that contain precursor chemicals that are specifically designed to be combined to produce homemade explosives.

Commercially available exploding target (ET) kits are pre-packaged consumer products consisting of precursor chemicals in the proper forms and ratio to make an HME, which are packaged with explicit instructions for how to make an HME that can be detonated in the absence of either a commercial detonator or any other primary explosives. ET kits contain either ammonium nitrate or potassium perchlorate as the oxidizer and powdered aluminum as the fuel, all three of which are Group A chemicals. Moreover, like other precursor chemicals under consideration in this report, ET kits are readily available at retail locations and online. ET kits were designed for target shooting; however, as the recent attacks in New York and New Jersey (see Chapter 2) demonstrate, ET kits can be used to construct IEDs.

Currently, no federal entity has explicit authority from Congress to oversee the sale of ET kits. As discussed in Chapter 3, the Bureau of Alcohol, Tobacco, Firearms and Explosives (ATF) regulates neither the distribution and sale nor the use of ET kits because the individual components, when unmixed, do not meet the definition of explosive materials, and when mixed after purchase, fall outside the scope of commerce in explosives that ATF enforces. The Department of Homeland Security (DHS) does not regulate the distribution and sale of ET kits, unless in limited circumstances where quantities exceed the Chemical Facilities Anti-Terrorism Standards (CFATS) thresholds at facilities where ET kits are manufactured, stored, or sold. A subset of states has implemented controls on the sale and use of ET kits. Some private entities have issued disclaimers about the use of their ET products.

The committee observes the following opportunities for federal, state, local, and private sector entities to take actions to address threats associated with ET kits:

- **U.S. DHS.** Congress could expand DHS's statutory authority to cover ET kits to enable DHS to explore strategies to reduce IED threats by overseeing the transactions of these kits to consumers.
- **U.S. ATF.** Congress could expand ATF's statutory definition of explosives to include ET kits, as several states have done, to create a common approach to security controls for explosives and ET kits.
- **State or Local Governments.** States or localities seeking to reduce threats associated with ETs may ban sales or institute licensing or permitting programs for the purchase or use of ET kits. They might also redefine terms such as explosives and lawful use so that they are covered under existing regulations.
- **Private Sector Entities.** Companies involved in the manufacture, distribution, and sale of ET kits could develop voluntary measures that reduce threats of misappropriation, for example, by securing materials, training employees to recognize suspicious behavior, and reporting suspicious behavior or transactions.

Analysis of Control Strategies

Recommendation 5: U.S. DHS should engage in a more comprehensive, detailed, and rigorous analysis of specific provisions for proposed mandatory and voluntary policy mechanisms to restrict access to precursor chemicals by malicious actors.

Consideration of potential control strategies should not end with this report. The committee began a process of assessment in Chapter 5 that can serve as a starting point for a closer examination of retail-level restrictions and other policy options. The committee conducted a qualitative assessment of retail-level control strategies that included a ban, licensing, a registry, existing controls, and various other measures and activities and could not identify a dominant option. Each type of control strategy would incur costs, benefits, and uncertainties that can only be weighed comprehensively with a more rigorous and more detailed analysis of specific policy mechanisms and a final evaluation of options. A relatively stringent option might, for example, do more to restrict malicious actors' access to precursor chemicals, yet entail more forgone sales and use and increased risks of displacement and commercial disruption, but the details will matter greatly to the analysis.

A full analysis, as would be necessary to support a policy decision, would require more time, data, industry participation, and specificity, including a clear articulation of the structure and content of proposed actions. Any such analysis should also consider the results of existing domestic programs that restrict access to precursor chemicals, including those intended to curb illicit drug production, and programs adopted in other countries. For example, the EU's experience with bans, licensing, and registries might provide valuable data for further analysis of those types of controls. As the EU's efforts mature, it might be possible to learn more about what is and is not working in the EU, the associated costs, and the applicability to the United States.

Similarly, DHS should use the available data on precursor chemical trends overseas to analyze new control systems, reevaluate priorities among precursor chemicals based on the threat level, and respond to new threats proactively.

The analysis recommended here should work to expand beyond the kind of breakeven analysis of benefits and costs that frequently informs executive branch decisions on security regulations (see Appendix G). Supplementary to the recommendation for further analysis, the committee urges the executive and legislative branches to identify and overcome impediments to such work, including time constraints and lack of funds, expertise, or suitable analytical tools and methods. Ideally, a more rigorous and more detailed analysis of options would enable executive and legislative branch officials to more deliberately consider a wide range of tradeoffs and would help build consensus about the benefits and costs of different control strategies.

Voluntary Measures, Activities, and Programs

Recommendation 6: The federal government should provide additional support for voluntary measures, activities, and programs that can contribute to restricting access by malicious actors to precursor chemicals used to manufacture IEDs.

Efforts to strengthen outreach, better educate retailers, and partner with commercial enterprises, including trade associations, in voluntary programs are likely to increase compliance with current security programs and can contribute positively in their own right. Anecdotal evidence suggests that commercial enterprises are willing to engage in voluntary programs tied to their memberships in trade groups, sponsored by trade groups, or sponsored by government agencies as private-public partnerships, or as outreach between the government and commercial enterprises.

ATF and FBI are actively trying to increase awareness and reporting of suspicious behavior. More specifically, ATF has focused on educating and developing relationships with the agribusiness industry to limit illegitimate access to precursor chemicals such as AN by increasing voluntary reporting of suspicious activity. FBI has engaged in efforts targeting retailers (e.g., pool and spa and beautician suppliers carrying hydrogen peroxide). To incentivize participation by commercial entities, government agencies might need to persuade them of the importance of the problem and necessity of voluntary action, or provide direct benefits or cost offsets, as might occur in a public-private partnership.

Existing program initiatives include *Be Aware for America* (1995), *Be Secure for America* (1998), and *America's Security Begins with You* (2004).

Trade group programs were reported to include a majority of businesses in certain industries (>90% in some cases), which encompasses thousands of individual locations related to manufacturing, chemical distribution, and agriculture.

Voluntary programs, such as those designed to educate retailers on how to identify suspicious activity, might be most effective if designed with input from businesses' leadership so that they can become part of the prevailing corporate culture. Some examples of programs that could be implemented either voluntarily or under mandates are listed below. Some trade groups, government agencies, or private-public partnerships already implement some of these options.

- **Developing outreach programs that target retailers**. Many retailers, either brick and mortar or online, that sell products that contain precursor chemicals may not be aware that these commercial products can be converted into explosive materials or contribute to the threat of IED attacks, as evidenced by the EU's experiences (see Chapter 4). To increase retailers' awareness of concerns and, hence, their ability to detect and report suspicious behavior and questionable sales transactions, trade associations or government agencies could create new outreach programs or strengthen existing ones. An outreach program could make use of a variety of educational tools, including online interactive programs, videos, pamphlets, in-person seminars, and other forms of information dissemination.
- **Training on programs and reporting of suspicious behavior, fraud, theft, and loss**. Whether policy makers choose to invoke a ban, licensing, a registry, or other mandatory or voluntary measures, retailers must have the knowledge and skills to implement the measures and report on suspicious behavior, fraud, theft, and loss. Trade associations, as some do currently, could develop tools, as listed above, and provide online or in-person training to support these goals, or government agencies could take the lead in development and implementation (see Chapters 3 and 4).

Training could cover techniques on requesting and verifying purchasing credentials, ascertaining the intended customer use, recognizing suspicious behavior, and reporting activity, such as purchasing a combination of precursor chemicals, to the designated authorities.

- **Maintaining documentation of transactions.** Documentation could include an electronic or written record of purchases that occur in a brick-and-mortar retail outlet or online, and data analytics that could, for example, correlate precursor chemicals, quantities sold, dates and times of purchases, and the purchaser's identity. This record of relevant information on sales transactions could assist authorities by providing actionable information to thwart an IED attack or to investigate one that has occurred.
- **Verifying training and program implementation.** Auditing, inspections, and other feedback mechanisms could be used to determine whether workers have been trained properly or if retailer outlets are adequately implementing program requirements. Trade associations and some federal agencies already conduct inspections, either announced or unannounced, on facilities that store precursor chemicals, and in the EU mystery shopping has been used to provide feedback to retailers and regulatory agencies on program implementation.

PRIORITY RESEARCH AREAS

In addition to the aforementioned recommendations, the committee identified several areas of research that could provide additional pathways for limiting access to precursor chemicals or designing appropriate policy mechanisms.

Data Collection from Incidents Involving Explosives

Detailed, verified, and comprehensive data on the precursor chemicals used in HMEs is required to track the evolution of the IED threat and to provide perspective for the analysis of effective measures to consider for regulatory action. The Department of Defense (DOD) has developed the Level 1 Explosive Ordnance Disposal Reporting Guidelines, providing an example of how an agency has collected tactical and technical information for its purposes related to explosives.[238] The committee suggests that the USBDC and other federal agencies focus on collecting detailed and verified data on incidents involving explosives. Focus areas include the following:

- how to improve the collection of relevant data on the chemical composition of the explosives used in IEDs and the precursor chemicals recovered during incidents and investigations;

- how the collected data may be verified when final laboratory results become available during the investigation;
- how to aggregate and present the data;
- how policy-making bodies can improve data entry compliance across bomb squads to ensure robust and timely collection; and
- how policy-making bodies can inform USBDC of their information requirements.

Substitute Chemicals

The likelihood of precursor chemicals being obtained from retail locations could be reduced if these chemicals were replaced by other chemicals that lack the utility to manufacture HMEs while providing the same legitimate service. For example, the AN in cold packs could be replaced by another nonprecursor salt that is sufficiently endothermic upon dissociation. While that example is relatively straightforward, other substitutions are not so obvious and would require additional research to identify, and may not be possible for chemicals such as oxidizers, which fulfill the same chemical role in HMEs and legitimate applications. Given the costs of research, development, and implementation, there would need to be clear incentives from the market or elsewhere to initiate the process.

Standardized Thresholds

As detailed in Chapters 3 and 4, a wide range of precursor chemical concentration and mass regulatory thresholds are used by various agencies, both domestic and international. The cited thresholds listed by the various agencies may create unnecessary confusion for legitimate commerce and may still not prevent malicious actors from purchasing the precursor chemical at the quantity or quality needed to manufacture HMEs. For example, while CFATS cites a hydrogen peroxide threshold concentration of 35%, lower compositions can be used to synthesize triacetone triperoxide (TATP), but the threshold exempts most retail-level products from oversight. A systematic study to identify the thresholds based on sound scientific principles will contribute toward the goal of preventing the use of precursor chemicals in HMEs.

Behavioral Responses

The committee suggests further research into behavioral responses of: (1) malicious actors, businesses, and end users to any proposed or implemented policy, and (2) the public and policy makers to the threat of a terrorist act or an actual terrorist act. As explored throughout the report (see Chapters 1, 2, 3, and 5, and Appendixes B and G), such behavioral responses can affect the efficacy of any strategy and the tradeoffs among strategies. Questions deserving attention include the following:

- What levels of knowledge, expertise, or ability would be required to circumvent a control that limits access to specific precursors?
- What level of real or perceived difficulty of obtaining a desired precursor will cause a terrorist to shift to a less desirable material or tactic, and will those tactics yield a greater or lesser overall risk?
- What levels of real and perceived risk and inconvenience are tolerated by business and the general public, and how would they respond to a new policy mechanism?
- How do we ingrain deliberative thinking processes into regulatory frameworks in such a way as to encourage appropriate consideration without forestalling decision making?

FINAL THOUGHTS

Precursor chemicals have played a major role and evolving role in the manufacture of HMEs in prior IED terrorist attacks. Given the ease of accessing precursor chemicals and the information required to manufacture HMEs, particularly via the internet, the threat of terrorist attacks using IEDs within the United States remains. While the United States has not experienced a major VBIED attack since the 1990s, the committee stresses the importance of engaging in deliberative thinking before crises to avoid the pitfalls of intuitive thinking and action bias during and after crises. In addition, the committee stresses the importance of periodically reevaluating and reprioritizing the precursor chemicals that it considered in this report and of building in means of harvesting and leveraging experience and data to strengthen policy over time. Restricting access to precursor chemicals may help to reduce the threat of IED attacks as terrorist tactics change, but only if policy can change, too.

The committee emphasizes the essential role of Congress in developing and implementing appropriate risk-reducing control strategies. Whether by addressing questions of federalism, defining the responsibilities of federal agencies, holding fact-finding hearings to articulate welfare interests and policy objectives, or adequately funding collaborative public-private work, Congress provides foundational opportunities for the executive branch to implement effective and efficient control strategies across administrations. Congress can be particularly instrumental in ensuring that crisis-driven interests do not unduly influence new laws or regulations, and can play a role in each of the six recommendations to enhance the nation's domestic and international risk-reduction programs.

References

1. UNMAS (United Nations Mine Action Service). 2016. *Improvised Explosive Device Lexicon*. New York: United Nations. http://www.mineaction.org/sites/default/files/publications/UNMAS%20IED%20Lexicon.pdf (accessed 06/18/2017).
2. NRC (National Research Council). 2007. *Countering the Threat of Improvised Explosive Devices: Basic Research Opportunities*. Washington, DC: The National Academies Press.
3. USBDC (U.S. Bomb Data Center). 2015. *Annual Explosives Incident Report*. https://www.atf.gov/rules-and-regulations/docs/report/2015usbdcexplosiveincidentreportpdf/download (accessed 9/21/2017).
4. CNN. 2017. Boston marathon terror attack fast facts. CNN Library: March 29, 2017. http://www.cnn.com/2013/06/03/us/boston-marathon-terror-attack-fast-facts (accessed 06/18/2017).
5. White House. 2013. *Countering Improvised Explosive Devices*. February 26, 2013. https://obamawhitehouse.archives.gov/sites/default/files/docs/cied_1.pdf (accessed 06/18/2017).
6. Stevens, G.D. 2012. *Whole of Government Approach to Countering IEDs: Leveraging Military Capabilities*. Institute for National Security and Counterterrorism, Syracuse University, Syracuse, NY. http://insct.syr.edu/wp-content/uploads/2013/02/Stevens_Whole-of-Government-Approach.pdf (accessed 06/18/2017).
7. Hallett, H.C. 2011. *Coroner's Inquests into the London Bombings of 7 July 2005*. H.M. Coroner's Office, London. http://webarchive.nationalarchives.gov.uk/20120216072447/http://7julyinquests.independent.gov.uk/docs/orders/rule43-report.pdf (accessed 06/18/2017).
8. BBC. 2015. Paris attacks: What happened on the night. News: December 9, 2015. http://www.bbc.com/news/world-europe-34818994 (accessed 06/08/2017).
9. BBC. 2016. Brussels explosions: What we know about airport and metro attacks. News: April 9, 2016. http://www.bbc.com/news/world-europe-35869985 (accessed 06/08/2017).
10. BBC. 2017. Manchester arena blast: 19 dead and more than 50 hurt. News: May 23, 2017. http://www.bbc.com/news/uk-england-manchester-40007886 (accessed 06/08/2017).
11. Wulf, D. 2016. Statement of Tasks and Intent of Sponsor. Presentation at the First Meeting on Reducing the Threat of Improvised Explosive Device Attacks by Restricting Access to Chemical Explosive Precursors, October 26, 2016, Washington, DC.

12. Wilson, C. 2007. Improvised Explosive Devices (IEDs) in Iraq and Afghanistan: Effects and Countermeasures. *Congressional Research Service Report* RS22330. https://fas.org/sgp/crs/weapons/RS22330.pdf (accessed 9/22//2017).
13. Antiterrorism and Effective Death Penalty Act 1996, Pub. L. No. 104-132, 110 Stat. 1214.
14. NRC. 1998. *Containing the Threat from Illegal Bombings: An Integrated National Strategy for Marking, Tagging, Rendering Inert, and Licensing Exposives and Their Precursors*. Washington, DC: National Academy Press.
15. EC (European Commission). 2010. *Summary of the Impact Assessment: Accompanying Document to the Regulation of the European Parliament and of the Council on the Marketing and Use of Explosives Precursors*. European Commission, Brussels, Belgium. http://ec.europa.eu/smart-regulation/impact/ia_carried_out/docs/ia_2010/sec_2010_1041_en.pdf (accessed 9/22//2017).
16. GAO (U.S. Government Accountability Office). 2012. *Combating Terrorism: State Should Enhance Its Performance Measures for Assessing Efforts in Pakistan to Counter Improvised Explosive Devices*. GAO-12-614. http://www.gao.gov/assets/600/590869.pdf (accessed 9/22//2017).
17. Gares, K.L., K.T. Hufziger, S.V. Bykov, and S.A. Asher. 2016. Review of explosive detection methodologies and the emergence of standoff deep UV resonance Raman. *J Raman Spectrosc* 47(1):124-141.
18. 6 CFR (Code of Federal Regulation) § 27. Chemical Facility Anti-Terrorism Standards (2017).
19. DHS (U.S. Department of Homeland Security). 2011. Ammonium Nitrate Security Program; Proposed rule. *Fed Reg* 76(149):46908-46957.
20. Consolidated Appropriations Act of 2007, Pub. L. No. 110-161, 121 Stat. 1844.
21. 6 U.S.C. (U.S. Code) § 488. Definitions (2016).
22. Bates, T. 1992. *Rads: The 1970 Bombing of the Army Math Research Center at the University of Wisconsin and Its Aftermath*. New York: HarperCollinsPublishers.
23. FBI (Federal Bureau of Investigation). 2017. Oklahoma City Bombing. https://www.fbi.gov/history/famous-cases/oklahoma-city-bombing (accessed 06/08/2017).
24. CBS. 2013. A look back at the Norway massacre. News: February 18, 2013. http://www.cbsnews.com/news/a-look-back-at-the-norway-massacre (accessed 06/08/2017).
25. U.K. House of Lord Debate: June 12, 1972, vol 331, cols 629-631.
26. Reuters. 2016. Turkey bans sale of some fertilisers after bomb attacks. News: June 9, 2016. http://www.reuters.com/article/us-turkey-security-fertilisers-idUSKCN0YV1MD (accessed 06/15/2017).
27. DHS. 2011. Ammonium Nitrate Security Program. *Fed Reg* 76(195):62311-62312.
28. EU (European Union). 2013. Regulation (EU) No. 98/2013 of the European Parliament and of the Council of January 15, 2013, on the marketing and use of explosives precursors. O.J. (L 39):1-11.
29. Kunreuther, H. 2017. Conceptual Frameworks for Dealing with Risk. Presentation at the Third Meeting on Reducing the Threat of Improvised Explosive Device Attacks by Restricting Access to Chemical Explosive Precursors, February 7, 2017, Washington, DC.
30. Kahneman, D. 2011. *Thinking, Fast and Slow*. New York: Farrar, Straus, and Giroux.
31. Sunstein, C.R., and R. Zeckhauser. 2011. Overraction to fearsom risks. *Environ Resour Econ* 48(3):435-449.
32. DA (U.S. Department of the Army). 2014. Risk Management. Pamphlet 385-30. U.S. Department of the Army, Washington, DC.
33. DA. 2014. Risk Management. ATP 5-19. http://www.benning.army.mil/RangeOps/content/blank_forms/ATP_5-19RiskManagement_Apr14.pdf (accessed 09/25/2017).
34. Kirk Yeager, FBI, Washington, DC, personal communication, April 19, 2017.
35. Cordeoy, J. 2014. *Material Harm*. London: Action on Armed Violence. https://aoav.org.uk/wp-content/uploads/2015/03/ied_material_lr.pdf (accessed 09/25/2017).
36. Rostberg, J.I. 2005. Common Chemicals as Precursors of Improvised Explosive Devices: The Challenges of Defeating Domestic Terrorism. *M.A. Thesis*, Naval Postgraduate School, Monterey, CA.

37. 27 CFR § 555.26. Prohibited Shipment, Transportation, Receipt, Possession, and Distribution of Explosive Materials (2017).
38. 27 CFR § 555.41. General (2017).
39. Spaaij, R. 2010. The enigma of lone wolf terrorism: An assessment. *Stud Confl Terror* 33(9):854-870.
40. Hayes, B.C., and I. McAllister. 2005. Public support for political violence and paramilitarism in Northern Ireland and the Republic of Ireland. *Terror Polit Violence* 17(4):599-617.
41. Beveridge, A., ed. 2011. *Forensic Investigation of Explosions*, 2nd Ed. Boca Raton, FL: CRC Press.
42. Foulger, B., and P. Hubbard. 1996. A review of techniques examined by UK authorities to prevent or inhibit the illegal use of fertiliser in terrorist devices. Pp. 129-133 in *Compendium of Papers of the International Explosives Symposium, September 18-22, 1995, Fairfax, VA*. Washington, DC: Bureau of Alcohol, Tobacco, Firearms and Explosives.
43. The President. 2011. *Executive Order* 13563. Improving Regulation and Regulatory Review, January 18, 2011. *Fed Reg* 76(14):3821-3823.
44. Vogen, G. 2016. CAN - Supply Chain Overview for NAS Committee. Presentation at the Second Meeting on Reducing the Threat of Improvised Explosive Device Attacks by Restricting Access to Chemical Explosive Precursors, December 13, 2016, Washington, DC.
45. Johnson, C.M. 2012. *U.S. Agencies Face Challenges Countering the Use of Improvised Explosive Devices in the Afghanistan/Pakistan Region*. GAO-12-907T. Washington, DC: U.S. Government Accountability Office. http://www.gao.gov/assets/600/592309.pdf (accessed 09/25/2017).
46. Liu, T. 2014. Modeling Continuous IED Supply Chains. *M.S. Thesis*, Air Force Institute of Technology, Wright-Patterson Air Force Base, OH.
47. Liu, T., and K. Pond. 2016. Modeling and estimating continuous improvised explosive device supply chain behavior. *J Def Model Simul Appl Methodol Technol* 13(1):67-75.
48. Conflict Armament Research. 2016. *Tracing the Supply of Components Used in Islamic State IEDs: Evidence from a 20-Month Investigation in Iraq and Syria*. http://www.conflictarm.com/wp-content/uploads/2016/02/Tracing_The_Supply_of_Components_Used_in_Islamic_State_IEDs.pdf (accessed 09/25/2017).
49. Conflict Armament Research. 2016. *Standardisation and Quality Control in Islamic State's Military Production: Weapon Manufacturing in the East Mosul Sector*. London: Conflict Armament Research.
50. Bevan, J. 2016. Tracing the Supply of Components Used in Islamic State IEDs. Presentation at the Second Meeting on Reducing the Threat of Improvised Explosive Device Attacks by Restricting Access to Chemical Explosive Precursors, December 13, 2016, Washington, DC.
51. Quihuis, N.R. 2012. Stemming the Flow of Improvised Explosive Device Making Materials through Global Export Control Regimes. *M.A. Thesis*, Naval Postgraduate School, Monterey, CA.
52. Vitasek, K. 2013. *Supply Chain Management Terms and Glossary*. Lombard, IL: Council of Supply Chain Management Professionals.
53. Sheffi, Y. 2005. *The Resilient Enterprise: Overcoming Vulnerability for Competative Advantage*. Cambridge, MA: The MIT Press.
54. Frazelle, E. 2002. *Supply Chain Strategy: The Logistics of Supply Chain Mannagement*. New York: McGraw-Hill.
55. Anderson, R. 2016. Hydrogen Peroxide and Sodium Chlorate. Presentation at the Second Meeting on Reducing the Threat of Improvised Explosive Device Attacks by Restricting Access to Chemical Explosive Precursors, December 13, 2016, Washington, DC.
56. Sattlethight, H. 2016. Presentation by the Aluminum Association on Aluminum Powder, Paste, and Flake at the Second Meeting on Reducing the Threat of Improvised Explosive Device Attacks by Restricting Access to Chemical Explosive Precursors, December 13, 2016, Washington, DC.

57. Hsu, N. 2016. Ammonium Nitrate - Its Manufacture and Supply Chain. Presentation at the Second Meeting on Reducing the Threat of Improvised Explosive Device Attacks by Restricting Access to Chemical Explosive Precursors, December 13, 2016, Washington, DC.
58. Liske, K. 2017. Agricultural Retailers and Chemical Explosive Precursors. Presentation at the Third Meeting on Reducing the Threat of Improvised Explosive Device Attacks by Restricting Access to Chemical Explosive Precursors, February 7, 2017, Washington, DC.
59. Lewis, M. 2017. Proactively Managing Supply Chain Safety and Security. Webinar on Reducing the Threat of Improvised Explosive Device Attacks by Restricting Access to Chemical Explosive Precursors, January 26, 2017.
60. Smith, P.E. 2010. *Teaching Chemistry with Pyrotechnic Flair: Laboratory Notebook.* Purdue University, West Lafayette, IN.
61. Paoli, G.P., J. Aldridge, N. Ryan, and R. Warnes. 2017. *Behind the Curtain: The Illicit Trade of Firearms, Explosives and Ammunition on the Dark Web.* Santa Monica, CA: RAND Corporation.
62. Sigma-Aldrich. 2017. Opening a Sigma-Aldrich Account. http://www.sigmaaldrich.com/site-level/corporate/ordering-support/new-account.html (accessed 06/08/2017).
63. Amazon. 2017. Dangerous Goods Identification Guide. https://www.amazon.com/gp/help/customer/display.html?ref=hp_rel_topic?ie=UTF8&nodeId=200339720 (accessed 06/08/2017).
64. eBay. 2017. Hazardous, Restricted, or Regulated Materials. https://pages.ebay.com/help/policies/hazardous-materials.html (accessed 06/08/2017).
65. eBay. 2017. Law Enforcement Center. https://pages.ebay.com/securitycenter/LawEnforcement-Center.html (accessed 06/08/2017).
66. Kennedy, R. 2016. Newcastle man who triggered police raid over explosives at his home was making home-made bangers. Chronicle Live News: October 3, 2016. http://www.chroniclelive.co.uk/news/north-east-news/newcastle-man-who-triggered-police-11969700 (accessed 06/08/2017).
67. Explosives Act of 1917, Pub. L. No. 65-68, 40 Stat. 385.
68. Federal Explosives Act of 1941, Pub. L. No. 77-381, 55 Stat. 863.
69. OIG (Office of the Inspector General). 2009. *Explosives Investigation Coordination between the Federal Bureau of Invesigation and the Bureau of Alcohol, Tobacco, Firearms and Explosives.* Audit Report 10-01. U.S. Department of Justice, Office of the Inspector General, Audit Division, Washington, DC. https://oig.justice.gov/reports/plus/a1001.pdf (accessed 06/08/2017).
70. 27 CFR § 555.23. List of Explosive Materials (2017).
71. ATF (Bureau of Alcohol, Tobacco, Firearms and Explosives). 2016. Commerce in explosives: Annual list of explosive materials. *Fed Reg* 81(221):80684-80686.
72. Moore Memorial Public Library. 2017. Texas City Disaster 1947. http://www.texascity-library.org/disaster/first.php (accessed 06/08/2017).
73. NRC. 1953. *A Compendium on the Hazards of Water Transportation and the Manufacture, Handling, Storage, and Stowage of Ammonium Nitrate and Ammonium Nitrate Fertilizers.* Washington, DC: National Academy Press.
74. Nature. 1921. The Oppau explosion. *Nature* 108(2713):278-279.
75. CSB (U.S. Chemical Safety and Hazard Investigation Board). 2013. *West Fertilizer Company Fire and Explosion.* Report 2013-02-I-TX. http://www.csb.gov/assets/1/19/West_Fertilizer_FINAL_Report_for_website_0223161.pdf (accessed 06/08/2017).
76. 27 CFR § 555.220. Table of Separation Distances of Ammonium Nitrate and Blasting Agents from Explosives or Blasting Agents (2017).
77. Guzman, A. 2017. Chemical Control Timeline: Brief History of U.S. Chemical Regulations. Presentation at the Third Meeting on Reducing the Threat of Improvised Explosive Device Attacks by Restricting Access to Chemical Explosive Precursors, February 8, 2017, Washington, DC.
78. 21 U.S.C. § 802. Definitions (2016).
79. DEA (Drug Enforcement Administration). 1999. Advisories to the Public: Special Surveillance List of Chemicals, Products, Materials and Equipment Used in the Clandestine Production of Controlled Substances. Diversion Control Division. https://www.deadiversion.usdoj.gov/chem_prog/advisories/surveillance.htm (accessed 06/13/2017).

80. INCB (International Narcotics Control Board). 2016. *Precursors and Chemicals Frequently Used in the Illicit Manufacture of Narcotic Drugs and Psychotropic Substances 2015*. New York: United Nations. https://www.incb.org/documents/PRECURSORS/TECHNICAL_REPORTS/2015/2015-PreAR_E.pdf (accessed 06/13/2017).
81. Domestic Chemical Diversion Control Act of 1993, Pub. L. No. 103-200, 107 Stat. 2333.
82. Comprehensive Methamphetamine Control Act of 1996, Pub. L. No. 104-237, 110 Stat. 3099.
83. Combat Methamphetamine Epidemic Act of 2005, Pub. L. No. 109-177, 120 Stat. 256.
84. Appriss Health. 2017. NPLEx. https://apprisshealth.com/solutions/nplex (accessed 07/06/2017).
85. 6 U.S.C. § 627. CFATS Regulations (2016).
86. Klessman, T., and K. Murray. 2017. Chemical Facility Anti-Terrorism Standards (CFATS) and Secure Handling of Ammonium Nitrate Overview. Presentation at the Third Meeting on Reducing the Threat of Improvised Explosive Device Attacks by Restricting Access to Chemical Explosive Precursors, February 7, 2017, Washington, DC.
87. IP (The Office of Infrastructure Protection). 2017. Chemical Facility Anti-Terrorism Standards (CFATS) Tiering Results Update. Webinar on CFATS Tiering Update - April 2017, April 24, 2017.
88. DHS. 2007. Appendix to Chemical Facility Anti-Terrorism Standards. *Fed Reg* 72(223): 65396-65435.
89. 6 CFR § 27.110. Chemical Facility Anti-Terorism Standards: Applicability (2017).
90. DHS. 2017. Tiering update. *CFATS Quarterly* 3(4).
91. DHS. 2017. Chemical Facility Anti-Terrorism Standards: Monthly Update. *Fact Sheet:* May 2017. https://www.dhs.gov/sites/default/files/publications/cfats-fact-sheet-05-17-508.pdf (accessed 07/06/2017).
92. Cooley, A., and B. McMenemy. 2017. USCG & MTSA Overview. Presentation at the Third Meeting on Reducing the Threat of Improvised Explosive Device Attacks by Restricting Access to Chemical Explosive Precursors, February 8, 2017, Washington, DC.
93. 46 U.S.C. § 70103. Maritame Transportation Security Plans (2016).
94. 33 CFR § 126. Handling of Dangerous Cargo at Waterfront Facilities (2017).
95. 46 CFR § 153.40. Determination of Materials That Are Hazardous (2016).
96. Tsoi, N., and D. Sindlinger. 2017. Hazardous Materials Endorsement Threat Assessment Program (HTAP) & Transportation Worker Identification Credential (TWIC). Presentation at the Fourth Meeting on Reducing the Threat of Improvised Explosive Device Attacks by Restricting Access to Chemical Explosive Precursors, April 19, 2017, Washington, DC.
97. 49 CFR § 1572. Credentialing and Security Threat Assessments (2016).
98. 15 U.S.C. § 2605. Prioritization, Risk Evaluation, and Regulation of Chemical Substances and Mixtures (2016).
99. 40 CFR §§ 700-799. Toxic Substances Control Act (2017).
100. EPA (U.S. Environmental Protection Agency). 2017. TSCA Requirements for Importing Chemicals https://www.epa.gov/tsca-import-export-requirements/tsca-requirements-importing-chemicals#specific (accessed 06/12/2017).
101. 42 U.S.C. §§ 11001-11050. Emergency Planning and Community Right to Know (2016).
102. 40 CFR § 372. Toxic Chemical Release Reporting: Community Right-to-Know (2017).
103. 29 CFR § 1910.109. Explosives and Blasting Agents (2017).
104. 29 CFR § 1910.119. Processs Safety Management of Highly Hazardous Chemicals (2017).
105. 29 CFR § 1910.38. Emergency Action Plan (2017).
106. ATF and MSHA (Bureau of Alcohol, Tobacco, Firearms and Explosives, and Mine Safety and Health Administration). 2016. Memorandum of Understanding between Department of Justice, Bureau of Alcohol, Tobacco, Firearms and Explosives and the Department of Labor, Mine Safety and Health Administration, June 24, 2016.
107. 49 CFR § 172.101. Purpose and Use of Hazardous Materials Table (2016).
108. 49 CFR § 172.102. Special Provisions (2016).

109. Webb, S. 2017. National Academy of Sciences Precursor Security. Presentation at the Third Meeting on Reducing the Threat of Improvised Explosive Device Attacks by Restricting Access to Chemical Explosive Precursors, February 8, 2017, Washington, DC.
110. Bomgardner, P. 2017. National Academy of Sciences (NAS) Security of Precursor Chemicals in Transportation. Presentation at the Third Meeting on Reducing the Threat of Improvised Explosive Device Attacks by Restricting Access to Chemical Explosive Precursors, February 8, 2017, Washington, DC.
111. 49 CFR § 385.415. What Operational Requirements Apply to the Transportation of a Hazardous Material for Which a Permit Required? (2016).
112. 49 CFR § 387.9. Financial Responsibility, Minimum Levels (2016).
113. 49 CFR § 172.800. Purpose and Applicability (2016).
114. 49 CFR § 1580. Rail Transportation Security (2016).
115. Holbein, J., S. Daniels, B.F. Elkins, J. Freas, R. Kane, J.W. Kitzmiller, V. Lee, D.G. Michels, L. Powell, D. Rimmer, D. Shepherdson, J.L. Summers, and C.S. Wilson. 2017. *Harmonized Tariff Schedule of the United States (2017) – Revision 1*. Washington, DC: U.S. International Trade Commission.
116. 15 CFR § 774. The Commerce Control List (2017).
117. BIS (Bureau of Industry and Security). 2016. *Commerce Control List*. U.S. Department of Commerce Bureau of Industry and Security, Washington, DC. https://www.bis.doc.gov/index.php/regulations/commerce-control-list-ccl (accessed 09/24/2017).
118. California Food and Agricultural Code §§ 14591-14593. Fertilizing Materials; Licensing (2016).
119. Illinois Fertilizer Act of 1961, 505 ILCS 80/12.
120. Iowa Code § 200.17A. Ammonium Nitrate Security (2016).
121. Kansas Statutes § 2-1201c. Ammonium Nitrate Dealers; Registration, Storage, Security, Records of Sale; Right to Refuse Sale (2016).
122. Maryland Code Agriculture § 6-209.1. Records of Sale or Distribution of Ammonium Nitrate Fertilizer (2016).
123. Michigan Compiled Laws § 324.8510. Inspecting, Sampling, and Analyzing Fertilizer and Soil Conditioners, Methods; Rules; Access to Premises; Stopping Conveyances (2016).
124. Montana Code §§ 80-10-201—80-10-212. Agriculture Commercial Fertilizers (2015).
125. Nevada Revised Statutes §§ 588.295-588.297. Commercial Fertilizers and Agricultural Minerals (2015).
126. New Jersey Revised Statutes § 4:9-15.43. Record of Sales of Restricted Commercial Fertilizer (2016).
127. New York Agriculture and Marketing Law §154. Ammonium Nitrate Security (2017).
128. Oklahoma Fertilizer Act and Rules § 35:30-29-37.1. Ammonium Nitrate Security (2017).
129. South Carolina Code of Laws § 46-25-210. Registration of Fertilizer; Application; Fee (2016).
130. Indiana Code § 15-16-2. Commercial Fertilizers (2016).
131. Texas Administrative Code § 65.6. Distribution of Ammonium Nitrate or Ammonium Nitrate Material (2017).
132. Wyoming Statutes § 11-14-118. Registration of Ammonium Nitrate (2016).
133. Davidson, P. 2007. Ammonium Nitrate Legislation by State. Presentation at the Third Meeting on Reducing the Threat of Improvised Explosive Device Attacks by Restricting Access to Chemical Explosive Precursors, February 8, 2017, Washington, DC.
134. Logue, C.A. 2017. New York State Department of Agriculture and Markets: Ammonium Nitrate Inspections. Presentation at the Fourth Meeting on Reducing the Threat of Improvised Explosive Device Attacks by Restricting Access to Chemical Explosive Precursors, April 19, 2017, Washington, DC.
135. Narrod, C. 2017. Public Private Partnership in International Food Safety Capacity Building. Presentation at the Third Meeting on Reducing the Threat of Improvised Explosive Device Attacks by Restricting Access to Chemical Explosive Precursors, February 7, 2017, Washington, DC.
136. 49 CFR § 1548.17. Known Shipper Program (2016).

137. OIG. 2009. *Transportation Security Administration's Known Shipper Program* (OIG-09-35). U.S. Department of Homeland Security Office of the Inspector General: Washington, DC. https://www.oig.dhs.gov/assets/Mgmt/OIGr_09-35_Mar09.pdf (accessed 09/26/2017).
138. 6 U.S.C. §§ 96 -973. Customs -Trade Partnership Against Terrorism (2016).
139. Krupinsky, S. 2017. C-TPAT. Presentation at the Third Meeting on Reducing the Threat of Improvised Explosive Device Attacks by Restricting Access to Chemical Explosive Precursors, February 8, 2017, Washington, DC.
140. Hilton, C. 2016. Institute of Makers of Explosives. Presentation at the Second Meeting on Reducing the Threat of Improvised Explosive Device Attacks by Restricting Access to Chemical Explosive Precursors, December 13, 2016, Washington, DC.
141. IME (Institute of Makers of Explosives). 2017. Safe Handling of Solid Ammonium Nitrate. *Safety Library Publication 30*. Washington, DC: Institute of Makers of Explosives.
142. IME. 2012. Security in Manufacturing, Transportation, Storage and Use of Commercial Explosives. *Safety Library Publication 27*. Washington, DC: Institute of Makers of Explosives.
143. IME. 2007. Recommendations for Accountability and Security of Bulk Explosives and Bulk Security Sensitive Materials. *Safety Library Publication 28*. Washington, DC: Institute of Makers of Explosives.
144. IATA (International Air Transport Association). 2017. Dangerous Goods. http://www.iata.org/whatwedo/cargo/dgr/Pages/index.aspx (accessed 06/13/2017).
145. Roczniak, D. 2017. American Chemistry Council Responsible Care & Security Code. Presentation at the Third Meeting on Reducing the Threat of Improvised Explosive Device Attacks by Restricting Access to Chemical Explosive Precursors, February 8, 2017, Washington, DC.
146. ACC (American Chemistry Council). 2017. Responsible Care. https://responsiblecare.americanchemistry.com (accessed 06/13/2017).
147. SOCMA (The Society of Chemical Manufacturers and Affiliates). 2017. ChemStewards Overview. http://www.socma.com/chemstewards (accessed 06/13/2017).
148. Gibson, J.C. 2016. Introduction to NACD and Responsible Distribution. Presentation at the Second Meeting on Reducing the Threat of Improvised Explosive Device Attacks by Restricting Access to Chemical Explosive Precursors, December 12, 2016, Washington, DC.
149. NACD (National Association of Chemical Distributors). 2017. About Responsible Distribution. https://www.nacd.com/rd/about (accessed 06/13/2017).
150. O'Hare, A. 2016. Improvised Explosives and the Fertilizer Industry. Presentation at the Second Meeting on Reducing the Threat of Improvised Explosive Device Attacks by Restricting Access to Chemical Explosive Precursors, December 12, 2016, Washington, DC.
151. ResponsibleAg. 2017. News. https://www.responsibleag.org (accessed 06/13/2017).
152. Yeager, K. 2011. What law enforcement needs to know about improvised explosives. *The Police Chief* LXXVIII(9):52-55. http://pages.nxtbook.com/nxtbooks/naylor/CPIM0911/offline/naylor_CPIM0911.pdf (accessed 10/1/2017).
153. McCrary, W, 2017. Bureau of Alcohol, Tobacco, Firearms and Explosives. Presentation at the Third Meeting on Reducing the Threat of Improvised Explosive Device Attacks by Restricting Access to Chemical Explosive Precursors, February 8, 2017, Washington, DC.
154. Sheehan, K., and M. Hendley. 2017. How We View the Threat. Presentation at the Third Meeting on Reducing the Threat of Improvised Explosive Device Attacks by Restricting Access to Chemical Explosive Precursors, February 8, 2017, Washington, DC.
155. DHS. 2015. *Catalog of Federally Sponsored Counter-IED Training and Education Resources for Private Sector Partners*. https://www.dhs.gov/sites/default/files/publications/obp-c-ied-training-resource-catalog-private-sector-508.pdf (accessed 09/28/2017).
156. Closs, D.J. 2016. Integrated Supply Chain Strategy. Presentation at the Second Meeting on Reducing the Threat of Improvised Explosive Device Attacks by Restricting Access to Chemical Explosive Precursors, December 12, 2016, Washington, DC.

157. DHS. 2012. *Chemical Sector Security Awareness Guide: A Guide for Owners, Operators, and Chemical Supply-Chain Professionals*. https://www.dhs.gov/sites/default/files/publications/DHS-Chemical-Sector-Security-Guide-Sept-2012-508.pdf (accessed 09/28/2017).
158. DHS. 2015. *Chemical Sector-Specific Plan: An Annex to the NIPP 2013*. https://www.dhs.gov/sites/default/files/publications/nipp-ssp-chemical-2015-508.pdf (accessed 09/28/2017).
159. Adams, J., P. Kurzer, A. Allen, P. Aubrey, E. Auner, R.G. Baird, C. Beecroft, N. Donohue, K.A. Grant, A. Kattan, and J.E. Nolan. 2013. *Remaking American Security: Supply Chain Vulnerabilities & National Security Risks Across the U.S. Defense Industrial Base*. Washington, DC: Alliance for American Manufacturing. http://docs.house.gov/meetings/FA/FA14/20130725/101216/HHRG-113-FA14-Wstate-AdamsB-20130725.pdf (accessed 09/28/2017).
160. Stanton, L. 2011. Theft and Diversion. Presentation at the Chemical Sector Security Summit, July 6, 2011, Baltimore, MD.
161. Cindrich, N. 2017. NAS Presentation on Chemical Precursor for IED. The Third Meeting on Reducing the Threat of Improvised Explosive Device Attacks by Restricting Access to Chemical Explosive Precursors, February 7, 2017, Washington, DC.
162. Ackerman, S. 3013. FBI warned in March that 'exploding targets' could fuel homemade bombs. News: April 17, 2013. https://www.wired.com/2013/04/tannerite (accessed 07/28/2017).
163. DOJ (U.S. Department of Justice). 2013. U.S. Forest Service Implements Closure Order to Prohibit Use of Exploding Targets on National Forest and Grasslands in Rocky Mountain Region, August 5, 2013. U.S. Attorney's Office District of Colorado. https://www.justice.gov/usao-co/pr/us-forest-service-implements-closure-order-prohibit-use-exploding-targets-national-forest (accessed 07/28/2017).
164. ATF. 2017. Binary Explosives. https://www.atf.gov/explosives/binary-explosives (accessed 07/26/2017).
165. 27 CFR § 555.11. Meaning of Terms (2017).
166. Tannerite. 2016. Legal Advisory. https://tannerite.com/legal-advisory (accessed 08/01/2017).
167. Rhode Island Senate Bill No. 2015-337.
168. NewYork Senate Bill No. S5276A.
169. Washington Revised Code § 70-74-022. License Required to Manufacture, Purchase, Sell, Use, Possess, Transport, or Store Explosive-Penalty (2017).
170. California Penal Code § 18720. Destructive Devices and Explosives Generally; Prohibited Acts (2016).
171. Massachusetts General Laws, Ch. 266 § 102. Possession or Control of Incendiary Device or Material; Possession of Hoax Device or Material Penalty (2016).
172. Maryland. Code Public Safety § 11-101. Definitions (2016).
173. Lousiana Laws Revised Statutes § 40:1472.2. Public Health and Safety Definitions (2016).
174. Code of Virginia § 18.2-85. Manufacture, Possession, Use, etc, of Fire Bombs or Explosive Materials or Devices; Penalties (2016).
175. Herring, M.R. 2014. Tannerite Advisory Opinion. Commonwealth of Virginia, Office of the Attorney General, Richmond, VA. https://www.vafire.com/content/uploads/2016/10/Enclosure-1-Attorney-General-Opinion-Tannerite-October-2014.pdf (accessed 09/29/17).
176. Oregon Revised Statutes § 480.210. Certificate, License or Permit Required (2015).
177. New Mexico Statutes § 30-7-19.1. Possession of Explosive Device or Incendiary Device (2016).
178. Colorado Revised Statutes § 18-12-109. Possession, Use, or Removal of Explosives or Incendiary Devices—Possession of Components Thereof—Chemical, Biological, and Nuclear Weapons—Persons Exempt—Hoaxes (2016).
179. Tennessee Code § 39-17-1301. Weapons; Part Definitions (2016).
180. Ohio Revised Code § 2923.17. Unlawful Possession of Dangerous Ordnance—Illegally Manufacturing or Processing Explosives (2016).
181. West Virginia Code § 61-3E-1. Offenses Involving Explosives; Definitions (2016).
182. New Jersey Revised Statutes § 21:1A-133. Permits for Manufacture, Sale, Storage, Transportation or Use of Explosives (2016).

183. Connecticut General Statutes § 29-349. Storage, Trasportation and Use of Explosives and Blasting Agents. Licenses, Permits: Fees, Suspension or Revocations (2015).
184. Hawaii Revised Statutes § 134-28. Explosive Devices; Prohibitions; Penalty (2016).
185. 25 Maine Revised Statutes § 2472. Explosives; Rules (2016).
186. Ohio State Fire Marshal. 2013. Guidance on Explosive Targets. https://www.geaugamapleleaf.com/wp-content/uploads/2014/11/sfm_guidance_on_explosive_targets_in_ohio.pdf (accessed 10/1/2017).
187. Slatery, H.H., A.S. Blumstein, and J.H. Wardle. 2015. Use of Exploding Targets. Opinion No. 15-12. State of Tennessee, Office of the Attorney General, Nashville, TN. https://www.tn.gov/assets/entities/attorneygeneral/opinions/op15-12.pdf (accessed 10/1/2017).
188. Chen, D. 2013. Acquiring and Possessing Explosives. OLR Research Report 2013-R-0222. Connecticut Office of Legislative Research. https://www.cga.ct.gov/2013/rpt/2013-R-0222.htm (accessed 10/1/2017).
189. Ostroskey, P.J. 2016. Binary Exploding Targets. Memorandum to Heads of Fire Departments, from State Fire Marshal, The Commonwealth of Massachusetts, Department of Fire Services. October 5, 2016.
190. Nebraska Revised Statutes § 28-1229. Explosives Control; Nebraska State Patrol; Permits; Issuance; Conditions (2016).
191. EC. 2017. *Report from the Commission to the European Parliament and the Council on the Application of, and Delegation of Power under Regulation (EU) 98/2013 of the European Parliament and of the Council on the Marketing and Use of Explosives Precursors*. European Commission, Brussels, Belgium.
192. Colley, T. 2017. International Precursor Regulations. Presentation at the Third Meeting on Reducing the Threat of Improvised Explosive Device Attacks by Restricting Access to Chemical Explosive Precursors, February 8, 2017, Washington, DC.
193. Australia Government. 2016. *National Code of Practice for Chemicals of Security Concern*. Attorney-General's Department Commonwealth of Australia, Canberra. https://www.nationalsecurity.gov.au/Securityandyourcommunity/ChemicalSecurity/Documents/Code-of-practice.PDF (accessed 10/1/2017).
194. Australian Government. 2016. Understanding the National Code of Practice for Chemicals of Security Concern Guide. https://www.nationalsecurity.gov.au/Securityandyourcommunity/ChemicalSecurity/Documents/Understanding-the-code-of-practice.pdf (accessed 10/1/2017).
195. PwC (PricewaterhouseCoopers). 2012. *Chemical Security: Precursors to Homemade Explosives—Decision Regulation Impact Statement*. http://ris.pmc.gov.au/sites/default/files/posts/2012/12/chemical-security-ris.pdf (accessed 10/1/2017).
196. Queensland Government. 2016. Regulation of Security Sensitive Ammonium Nitrate (SSAN) in Queensland. https://www.dnrm.qld.gov.au/mining/safety-and-health/alerts-bulletins-search1/alerts-bulletins/explosives/regulation-of-security-sensitive-ammonium-nitrate-ssan-in-queensland (accessed 07/10/2017).
197. Government of South Australia. 2017. License Conditions SSAN: Security Sensitive Ammonium Nitrate. *Technical Note 57*. https://www.safework.sa.gov.au/uploaded_files/SSAN_LicCondsT57.pdf (accessed 07/10/2017).
198. Australian Government. 2015. Telecommunications (Interception and Access) Amendment (Data Retention) Act No. 39.
199. Australian Government. 2014. Counter-Terrorism Legislation Amendment (Foreign Fighters) Act No. 116.
200. Australian Government. 2016. Counter-Terrorism Legislation Amendment Act No. 1, and Act No. 82.
201. Australian Government. 2015. *Review of Australia's Counter-Terrorism Machinery*. Department of the Prime Minister and Cabinet. https://www.pmc.gov.au/sites/default/files/publications/190215_CT_Review_1.pdf (accessed 07/12/2017).

202. Phan, S., A. McNeilage, and M. Levy. 2014. Live: Anti-terrorism raids across Sydney and Brisbane. The Sydney Morning Herald: September 18, 2014. http://www.smh.com.au/national/live-antiterrorism-raids-across-sydney-and-brisbane-20140917-3fzkq.html (accessed 07/12/2017).
203. ABC. 2014. Anti-terror raids in Melbourne: Man charged with sending funds to Islamic State and Al Qaeda affiliate. News: September 30, 2014. http://www.abc.net.au/news/2014-09-30/counter-terrorism-raids-afp-and-victoria-police-melbourne/5778116 (accessed 07/12/2017).
204. Australian Security Magazine. 2015. Two arrested at Fairfield on terrorism offences. News: February 12, 2015. https://australiansecuritymagazine.com.au/two-arrested-fairfield-terrorism-offences (accessed 07/12/2017).
205. Mark, D. 2015. Police say they've foiled murderous terror attack. ABC News: February 11, 2015. http://www.abc.net.au/news/2015-02-11/police-say-theyve-foiled-murderous-terror-attack/6086842 (accessed 07/12/2017).
206. Rubinsztein-Dunlop, S. 2015. Authorities warn of 'new normal' after thwarting terrorist plot. ABC News: February 11, 2015. http://www.abc.net.au/7.30/content/2015/s4178317.htm (accessed 07/12/2016).
207. Government of Canada. 2013. Explosives Regulations SOR/2013-211.
208. Natural Resources Canada. 2013. Evaluation of the Explosives Safety and Security Branch (ESSB) of the Minerals and Metals Sector (MMS). https://www.nrcan.gc.ca/evaluation/reports/2010/834#c3b (accessed 06/13/2017).
209. Toronto Metro. 2017. List of some terrorism cases in Canada. News: January 31, 2017. http://www.metronews.ca/news/toronto/2017/01/31/list-of-some-terrorism-cases-in-canada.html (accessed 07/12/2017).
210. Lam, L. 2017. Singapore under highest terror threat in recent years: 8 key points from MHA's terror report. The Straits Times, June 1, 2017. http://www.straitstimes.com/singapore/singapore-under-highest-terror-threat-in-recent-years-8-key-points-from-mhas-terror-report (accessed 07/18/2017).
211. Singapore Government. 2017. *Singapore Terrorism Threat Assessment Report 2017*. Ministry of Home Affairs. https://www.mha.gov.sg/newsroom/press-releases/Pages/Singapore-Terrorism-Threat-Assessment-Report-2017.aspx (accessed 07/18/2017).
212. CTTSO (Combating Terrorism Technical Support Office). 2017. International Partners. http://www.tswg.gov/?q=international_partners (accessed 07/18/2017).
213. Junn, L.C., and C. Oh. 2016. Simulated terror attack at Esplanade for Exercise Heartbeat 2016. Channel NewsAsia, September 2, 2016. http://www.channelnewsasia.com/news/singapore/simulated-terror-attack-at-esplanade-for-exercise-heartbeat-2016-7818742 (accessed 07/18/2017).
214. Singapore Government. 2003. Arms and Explosives Act, Chapter 13, Section 21. Licensing of Explosive Precursors.
215. bin Ali, A., and L.K. Yin. 2007. New Control Measures for 15 Chemicals in Singapore. Memorandum to All Affected HS Licenses of NE. Singapore Government, National Environment Agency, June 26, 2007. http://www.nea.gov.sg/docs/default-source/anti-pollution-radiation-protection/chemical-pollution/new-control-measures-for-15-chemicals-(explosive-precursors)-in-singapore.pdf (accessed 10/4/2017).
216. COREPER (Committee of Permanent Representatives in the European Union). 2008. EU Action Plan on Enhancing the Security of Explosives (8311/08). Note to Council of the European Union, Brussels, Belgium, April 11, 2008. http://register.consilium.europa.eu/doc/srv?l=EN&f=ST%208311%202008%20INIT (accessed 10/4/2017).
217. EC. 2017. Standing Committee on Precursors (E03245). http://ec.europa.eu/transparency/regexpert/index.cfm?do=groupDetail.groupDetail&groupID=3245 (accessed 06/13/2017).
218. EC. 2017. Migration and Home Affairs—Explosives precursors. https://ec.europa.eu/home-affairs/what-we-do/policies/crisis-and-terrorism/explosives/explosives-precursors_en (accessed 06/13/2016).
219. EU. 2008. Consolidated version of the Treaty on the Functioning of the European Union, art. 258. O.J. EU 51(C 115):47.

REFERENCES

220. EC. 2017. List of Measures to Implement Regulation (EU) 98/2013 on Explosives Precursors in the EU/EEA. European Commission, Brussels, Belgium. https://ec.europa.eu/home-affairs/sites/homeaffairs/files/what-we-do/policies/crisis-and-terrorism/explosives/explosives-precursors/docs/list_of_measures_en.pdf (accessed 10/4/2017).
221. Statutory Rules of Northern Ireland. 2014. The Control of Explosives Precursors etc. Regulations (Northern Ireland), SR 2014/224.
222. EU Regulation on the Marketing and Use of Explosives Precursors (2014).
223. Nagesh, A. 2017. Five terror plots thwarted since Westminster attack in March. Metro News: May 25, 2017. http://metro.co.uk/2017/05/25/five-terror-plots-thwarted-since-westminster-attack-in-march-6661876 (accessed 07/12/2017).
224. UK Government. 2017. Action Counters Terrorism: Reporting Terrorist or Extremist Content Online. https://act.campaign.gov.uk (accessed 06/26/2017).
225. Humberside Police. 2017. Court Case of Note – Gert Meyer. News: March 29, 2017. http://www.humberside.police.uk/news/court-case-note-%E2%80%93-gert-meyer (accessed 06/26/2017).
226. WCO (World Customs Organization). 2013. *Programme Global Shield*. http://www.wcoomd.org/-/media/wco/public/global/pdf/topics/enforcement-and-compliance/activities-and-programmes/security-programme/pgs/programme-global-shield-en.pdf?db=web (accessed 10/4/2017).
227. Goodman, M.B. 2015. *An Ounce of Prevention: Building a Global Shield to Defeat Improvised Explosive Devices*. Washington, DC: Center for American Progress. https://www.americanprogress.org/issues/security/reports/2015/06/04/114466/an-ounce-of-prevention/ (accessed 10/4/2017).
228. Woods, J.P. 2012. Testimony for a House Committee on Homeland Security, Subcommittee on Cybersecurity, Infrastructure Protection, and Security Technologies Hearing "Securing Ammonium Nitrate: Using Lessons Learned in Afghanistan to Protect the Homeland from IEDs." July 12, 2012. https://www.dhs.gov/news/2012/07/12/written-testimony-ice-hsi-nsi-assistant-director-john-woods-closed-hearing-titled (accessed 10/4/2017).
229. AOAV (Action on Armed Violence). 2016. Programme Global Shield. https://aoav.org.uk/2016/programme-global-shield (accessed 07/18/2017).
230. WCO. 2016. *Illicit Trade Report 2015*. World Customs Organization, Brussels, Belgium. http://www.wcoomd.org/-/media/wco/public/global/pdf/topics/enforcement-and-compliance/activities-and-programmes/illicit-trade-report/itr-2015-en.pdf?db=web (accessed 10/4/2017).
231. OMB (Office of Management and Budget). 2003. Circular A-4, September 17, 2003. https://www.whitehouse.gov/sites/whitehouse.gov/files/omb/circulars/A4/a-4.pdf (accessed 10/4/2017).
232. Langemeier, M. 2015. What Is Risk and Uncertinaty? Center for Commercial Agriculture, Purdue University. https://ag.purdue.edu/commercialag/Documents/Resources/Risk/Introduction/2015_01_Langemeier_Risk_Uncertainty.pdf (accessed 10/4/2017).
233. Kliesen, K.L. 2013. Uncertainty and the economy. *Regional Economist* 21(2).
234. AmosWEB. 2017. Economics of Uncertainty. http://www.amosweb.com/cgi-bin/awb_nav.pl?s=wpd&c=dsp&k=economics+of+uncertainty (accessed 07/13/2017).
235. Sen, A. 2009. *The Idea of Justice*. Cambridge, MA: Belknap Press.
236. Greenfield, V.A., H.H. Willis, and T. LaTourrette. 2012. *Assessing the Benefits of U.S. Customs and Border Protection Regulatory Actions to Reduce Terrorism Risks*. Santa Monica, CA: RAND Corporation.
237. Farrow, S., and S. Shapiro. 2009. The benefit-cost analysis of security focused regulations management. *J Homel Secur Emerg* 6(1):Art. 25.
238. Chairman of the Joint Chiefs of Staff, Joint Explosive Ordnance Disposal. 2016. *Joint Publication 3-42*.
239. DA. 1998. *Risk Management (FM 100-14)*. Department of the Army, Washington, DC. April 23, 1998.

240. DHS. 2011. *Risk Management Fundamentals: Homeland Security Risk Management Doctrine.* U.S. Department of Homeland Security, Washington, DC. https://www.dhs.gov/xlibrary/assets/rma-risk-management-fundamentals.pdf (accessed 07/13/2017).
241. FAA (Federal Aviation Administration). 2009. *Risk Management Handbook (FAA-H-8083-2).* U.S. Department of Transportation, Federal Aviation Administration, Washington, DC. https://www.faa.gov/regulations_policies/handbooks_manuals/aviation.
242. DHS. 2010. *DHS Risk Lexicon 2010 Edition.* Risk Steering Committee. https://www.dhs.gov/xlibrary/assets/dhs-risk-lexicon-2010.pdf (accessed 10/4/2017).
243. Cox, L.A. 2008. What's wrong with risk matrices? *Risk Anal* 28(2):497-512.
244. Rozell, D.J. 2015. A cautionary note on qualitative risk ranking of homeland security threats. *HSAJ* 11:Art. 3.
245. NRC. 2010. *Review of the Department of Homeland Security's Approach to Risk Analysis.* Washington, DC: The National Academies Press.
246. Lundberg, R., and H. Willis. 2015. Assessing Homeland Security risks: A comparative risk assessment of ten hazards. *HSAJ* 11:Art. 10.
247. Greenfield, V.A., and F. Camm. 2005. *Risk Management and Performance in the Balkans Support Contract.* Santa Monica, CA: RAND Corporation. https://www.rand.org/content/dam/rand/pubs/monographs/2005/RAND_MG282.sum.pdf (accessed 10/5/2017).
248. Greenfield, V.A., and L. Paoli. 2012. If supply oriented drug policy is broken, can harm reduction help fix it? Melding disciplines and methods to advance international drug control policy. *Int J Drug Policy* 23(1):6-15.
249. Amazon. 2017. 1 lb aluminum powder 30 micron 500 mesh by ESKS (spoon not included). https://www.amazon.com/Aluminum-Powder-Micron-Mesh-ESKS/dp/B06ZYP1VM6 (accessed 06/14/2017).
250. The Aluminum Association. 2016. Shipments of Aluminum Powder and Paste as Reported by Participating Companies. Industry Statistics, The Aluminum Association, Arlington, VA.
251. EPA (U.S. Environmental Protection Agency). 2015. *List of Lists: Consolidated List of Chemicals Subject to the Emergency Planning and Community Right-To-Know Act (EPCRA), Comprehensive Environmental Response, Compensation and Liability Act (CERCLA) and Section 112(r) of the Clean Air Act.* EPA 550-B-15-001. Office of Solid Waste and Emergency Response. https://www.epa.gov/sites/production/files/2015-03/documents/list_of_lists.pdf (accessed 06/14/2017).
252. EPA. 2017. TSCA Chemical Substance Inventory. https://www.epa.gov/tsca-inventory (accessed 07/12/2017).
253. PHMSA (Pipeline and Hazardous Materials Safety Administration). 2017. Hazmat Table. https://www.phmsa.dot.gov/staticfiles/PHMSA/DownloadableFiles/Files/Hazmat/Alpha_Hazmat_Table.pdf (accessed 07/12/2017).
254. Thomas, D. 2016. Overview of the Control Measures for the Domestic Supply Chain for Ammonium Nitrate and Nitric Acid. Presentation at the Second Meeting on Reducing the Threat of Improvised Explosive Device Attacks by Restricting Access to Chemical Explosive Precursors, December 13, 2016, Washington, DC.
255. DHS. 2016. Research on Supply and Uses of Improvised Explosive Device Precursors (IEDPs), December 9, 2016. Infrastructure Security Compliance Division, U.S. Department of Homeland Security, Washington, DC.
256. Sigma-Aldrich. 2017. Ammonium nitrate 99.999% trace metals basis. http://www.sigmaaldrich.com/catalog/product/aldrich/256064?lang=en&region=US (accessed 06/14/2017).
257. Amazon. 2017. Primacare PCP69 instant cold pack with cover size 6" x 9" (pack of 24). https://www.amazon.com/Primacare-PCP-69-Instant-Cold-Cover/dp/B00BEEC3NS (accessed 06/14/2017).
258. Amazon. 2017. Non-coated ammonium nitrate 34-0-0 prill form fertilizer - 25 lb. https://www.amazon.com/Non-Coated-Ammonium-Nitrate-34-0-0-Fertilizer/dp/B00O6EY5BG (accessed 06/14/2017).

REFERENCES

259. Tannerite. 2017. Single case of 1 pounders. https://tannerite.com/product/tannerite-single-case-of-1-pounders (accessed 06/14/2017).
260. Tannerite. 2017. Exploding rifle targets. Bass Pro Shops. http://www.basspro.com/Tannerite-Exploding-Rifle-Targets/product/1410170741566 (accessed 06/14/2017).
261. Amazon. 2017. Dynarex instant cold pack, 5 inches x 9 inches, 24 count. https://www.amazon.com/Dynarex-Instant-Cold-Inches-24-Count/dp/B004CQ5NUK (accessed 06/14/2017).
262. MKM Pool Spa. 2017. Aqua silk chlorine-free shock oxidizer 4x1 gal. https://www.mkmpoolspa.com/aqua-silk-chlorine-free-shock-oxidizer-4x1-gal/ (accessed 06/14/2017).
263. CVS. 2017. O W & CO hydrogen peroxide 12%, 16 oz. http://www.cvs.com/shop/health-medicine/first-aid/antibiotic-antiseptic/o-w-co-hydrogen-peroxide-12-16-oz-prodid-785437 (accessed 06/14/2017).
264. Fisher Scientific. 2017. Hydrogen peroxide 60% (w/v) (200 volumes), extra pure SLR. https://www.fishersci.co.uk/shop/products/hydrogen-peroxide-60-w-v-200-volumes-extra-pure-slr/10336280 (accessed 06/14/2017).
265. Amazon. 2017. 35% food grade hydrogen peroxide. https://www.amazon.com/35-Food-Grade-Hydrogen-Peroxide/dp/B0050DS18U (accessed 06/14/2017).
266. Arkema. 2005. *Hydrogen Peroxide: Brighten Your Future with Us.* http://americas.brightenyourfuture.com/export/sites/byf-americas/.content/medias/downloads/literature/hydrogen-peroxide-brochure.pdf (accessed 06/14/2017).
267. Nexant. 2008. *Hydrogen Peroxide: Process Technology, Production Costs (including Economic Comparison of Solvay's Mega-Scale Technology, Emerging Direct Synthesis and Conventional Anthraquinone Routes), and Regional Supply/Demand Forecasts Are Presented. PERP07/08-3.* ChemSystems PERP Program.
268. Amazon. 2017. Ultimate silver & gold testing kit nitric acid test solutions, stone, and jewelry scale. https://www.amazon.com/Ultimate-Silver-Testing-Solutions-Jewelry/dp/B00IN93TEE (accessed 06/14/2017).
269. Sigma-Aldrich. 2017. Nitric acid fuming, 99.5%. http://www.sigmaaldrich.com/catalog/product/vetec/v001092?lang=en&region=US (accessed 06/14/2017).
270. MRP Motorsports. 2017. TORCO racing fuel nitro methane drum. http://www.mrpmotorsports.net/ecommerce/torco-racing-fuels/torco-nitro-methane-drum.jsp (accessed 06/14/2017).
271. Walmart. 2017. Redcat racing nitrofuel nitro fuel. https://www.walmart.com/ip/Redcat-Racing-Nitrofuel-Nitro-Fuel/24901790 (accessed 06/14/2017).
272. Sigma-Aldrich. 2017. Nitromethane, ACS reagent, ≥95%, CH_3NO_2. http://www.sigmaaldrich.com/catalog/product/sigald/360554?lang=en&region=US (accessed 06/14/2017).
273. Fisher Scientific. 2017. Potassium chlorate (crystalline/certified ACS). https://www.fishersci.com/shop/products/potassium-chlorate-crystalline-certified-acs-fisher-chemical-2/p-32883 (accessed 06/14/2017).
274. Pyro Chem Source. 2017. Potassium chlorate. http://www.pyrochemsource.com/Potassium Chlorate-POT CHLOR.htm;jsessionid=3C4C1E86C1D7A401706A78722673E7E6.p3plqscs-fapp004 (accessed 06/14/2017).
275. Griffith, K., and K. Richmond. 2017. Potassium Perchlorate. Webinar on Reducing the Threat of Improvised Explosive Device Attacks by Restricting Access to Chemical Explosive Precursors, January 26, 2017.
276. Heckman, J.L. 2016. American Pyrotechnics Association. Presentation at the Second meeting on Reducing the Threat of Improvised Explosive Device Attacks by Restricting Access to Chemical Explosive Precursors, December 12, 2016, Washington, DC.
277. Rakuten. 2017. Potassium perchlorate high purity. https://www.rakuten.com/prod/potassium-perchlorate-high-purity/287638239.html (accessed 06/14/2017).
278. Sigma-Aldrich. 2017. Potassium perchlorate ≥99.99% trace metals basis. http://www.sigmaaldrich.com/catalog/product/aldrich/460494?lang=en&region=US (accessed 06/14/2017).

279. Amazon. 2017. 2 lb sodium chlorate crystals technical grade NaClO$_3$ weed killer. https://www.amazon.com/Sodium-Chlorate-Crystals-Technical-NaClO3/dp/B06XZTTZC3 (accessed 06/14/2017).
280. Fisher Scientific. 2017. Sodium chlorate (crystalline/certified), Fisher chemical glass bottle. https://www.fishersci.com/shop/products/sodium-chlorate-crystalline-certified-fisher-chemical-glass-bottle-500g/s268500 (accessed 06/14/2017).
281. Merchant Research & Consulting, Ltd. 2014. Global sodium chlorate production to reach 4 mln tonnes in 2017. News: June 20, 2014. https://mcgroup.co.uk/news/20140620/global-sodium-chlorate-production-reach-4-mln-tonnes.html (accessed 11/22/2016).
282. Plant Food Company Inc. 2017. 30-0-0, UAN Green-T. https://www.plantfoodco.com/golf-professional-turf/products/liquid-fertilizer/30-0-0-uan/ (accessed 07/12/2017).
283. Swedish Civil Contingencies Agency, Home Office, National Coordinator for Security and Counterterrorism, and Norwegian Directorate for Civil Protection. 2016. You Can Make a Difference! - Sell Chemical Products Responsibly. http://www.mh.government.bg/media/filer_public/2016/02/09/sos_leaflet_en.pdf (accessed 10/5/2017).
284. The President. 1993. Executive Order 12866 of September 30, 1993: Regulatory planning and review. *Fed Reg* 58 (190).
285. DHS. 2009. *National Infrastructure Protection Plan: Partnering to Enhance Protection and Resiliency.* https://www.dhs.gov/sites/default/files/publications/national-infrastructure-protection-plan-2009-508.pdf (accessed 10/6/2017).
286. Willis, H.H., A.R. Morral, T.K. Kelly, and J.J. Medby. 2005. *Estimating Terrorism Risk.* Santa Monica, CA: RAND Corporation.
287. Smith, V.K., C. Mansfield, and L. Clayton. 2009. Valuing a Homeland Security policy: Countermeasures for the threats from shoulder mounted missiles. *J Risk Uncertainty* 38(3):215-243.
288. Willis, H.H., and T. LaTourrette. 2008. Using probabilistic terrorism risk modeling for regulatory benefit-cost analysis: Application to the Western Hemisphere Travel Initiative in the land environment. *Risk Anal* 28(2):325-339.
289. Farrow, S. 2007. The economics of Homeland Security expenditures: Foundational expected cost-effectiveness approaches. *Contemp Econ Policy* 25(1):14-26.
290. Ezell, B.C., S.P. Bennett, D. von Winterfeldt, J. Sokolowski, and A.J. Collins. 2010. Probabilistic risk analysis and terrorism risk. *Risk Anal* 30(4):575-589.
291. Merrick, J., and G.S. Parnell. 2011. A comparative analysis of PRA and intelligent adversary methods for counterterrorism risk management. *Risk Anal* 31(9):1488-1510.
292. Jackson, B.A., P. Chalk, R.K. Cragin, B. Newsome, J.V. Parachini, W. Rosenau, E.M. Simpson, M. Sisson, and D. Temple. 2007. *Breaching the Fortress Wall: Understanding Terrorist Efforts to Overcome Defensive Technologies.* Santa Monica, CA: RAND Corporation. https://www.rand.org/content/dam/rand/pubs/monographs/2007/RAND_MG481.pdf (accessed 10/6/2017).
293. Morral, A.R., and B.A. Jackson. 2009. *Understanding the Role of Deterrence in Counterterrorism Security.* Santa Monica, CA: RAND Corporation. https://www.rand.org/pubs/occasional_papers/OP281.readonline.html (accessed 10/6/2017).

Appendix A

Acronyms

ACC	American Chemistry Council
AN	ammonium nitrate
AN/FO	ammonium nitrate / fuel oil
ANNIE	ammonium nitrate nitrobenzene improvised explosive
ANSP	Ammonium Nitrate Security Program
ATF	Bureau of Alcohol, Tobacco, Firearms and Explosives
BAU+	business as usual plus
BCA	benefit-cost analysis
BP	black powder
CAN	calcium ammonium nitrate
CBP	U.S. Customs and Border Protection
CDL	commercial drivers' license
CERCLA	Comprehensive Environmental Response, Compensation, and Liability Act
CFATS	Chemical Facility Anti-Terrorism Standards
CHP	concentrated hydrogen peroxide
COAG	Council of Australian Governments
C-TPAT	Customs-Trade Partnership Against Terrorism
DA	Department of the Army
DEA	Drug Enforcement Administration
DHS	U.S. Department of Homeland Security
DOC	U.S. Department of Commerce

DOD U.S. Department of Defense
DOJ U.S. Department of Justice
DOL U.S. Department of Labor
DOT U.S. Department of Transportation

EAP emergency action plan
EC European Commission
EGDN ethylene glycol dinitrate
EPA U.S. Environmental Protection Agency
EPCRA Emergency Planning and Community Right-to-Know Act
ERD Explosives Regulatory Division
ET exploding target
EU European Union
EU MS European Union member states

FALN Fuerzas Armadas de Liberación Nacional
FARC Fuerzas Armadas Revolucionarias de Colombia
FBI Federal Bureau of Investigation
FDA U.S. Food and Drug Administration
FGAN fertilizer grade ammonium nitrate

GPS global positioning system

HME homemade explosive
HMTD hexamethylene triperoxide diamine

IATA International Air Transport Association
ID identification
IED improvised explosive device
IME Institute of Makers of Explosives
IS icing sugar
ISIS Islamic State

KSP Known Shipper Program

MS member states (of the EU)
MSHA Mine Safety and Health Administration
MTSA Maritime Transportation Security Act

NACD National Association of Chemical Distributors
NFPA National Fire Protection Association
NG nitroglycerine
NPPD Nation Protection and Programs Directorate

APPENDIX A

OMB	Office of Management and Budget
OSHA	Occupational Safety and Health Administration
PBIED	person-borne improvised explosive device
PETN	pentaerythritol tetranitrate
PGS	Programme Global Shield
PHMSA	Pipeline and Hazardous Materials Safety Administration
PIRA	Provisional Irish Republican Army
PPF	powder, paste, and flake
SCP	Standing Committee on Precursors
SOCMA	Society of Chemical Manufacturers and Affiliates
SSAN	security sensitive ammonium nitrate
TATP	triacetone triperoxide
TGAN	technical grade ammonium nitrate
TNT	trinitrotoluene
TSA	Transportation Security Administration
TSCA	Toxic Substances Control Act
TWIC	Transportation Workers Identification Credential
UAN	urea ammonium nitrate
UFF	United Freedom Front
UN	United Nations
UPS	United Parcel Service
U.S.	United States
USBDC	U.S. Bomb Data Center
USCG	U.S. Coast Guard
USPS	U.S. Postal Service
VBIED	vehicle-borne improvised explosive device

Appendix B

Risk and Risk Management

In this report the committee has adopted the conceptual foundation and nomenclature of the national-security community and defined risk in terms of a threat or hazard, the bad consequences that can arise from that condition, and the probability and severity of those consequences. Specifically, the committee has worked with definitions from national-security doctrine that are broadly applicable with minor adaptations.[32,33] These definitions include the following:

- *Risk*: probability and severity of loss linked to hazards.
- *Hazard*: a condition with the potential to cause injury, illness, or death of personnel; damage to or loss of equipment or property; or mission degradation.
- *Probability*: the likelihood that an event will occur in an exposure interval (time, area, etc.), ranging from frequent or inevitable to unlikely or improbable.
- *Severity*: the expected consequences of an event in terms of injury, property damage, or other mission-impairing factors, ranging from catastrophic to negligible.

The committee has drawn from this doctrine, but paraphrased and adapted it for the report's purposes.

This doctrine dates back to the 1980s[239] and largely parallels the guidance of other federal agencies, including those with civilian and strategic concerns, for example, the Department of Homeland Security (DHS)[240] and the Department of Transportation.[241] DHS defines risk similarly as the "potential for an unwanted outcome resulting from an incident, event, or occurrence, as determined by its likelihood and the associated consequences."[242] The committee chose to work with the national-security doctrine because of its maturity and utility. It pinpoints two concrete facets of risk—probability and severity—and ties them to a set of practical risk-management tools that can support qualitative analysis. Those tools are not without criticism, which the committee addresses in a later note in this appendix.[243,244] For a review of the strengths and weaknesses of DHS's approach to risk analysis, see the cited literature.[245]

Two aspects of the committee's use of this vocabulary merit attention. First, whereas the national-security doctrine focuses on personnel and mission, the committee has adopted a broader lens that encompasses individuals, public and private institutions, and the environment. Second, the doctrine acknowledges the potential for losses from hazards (accidental), threats (tactical), acts of terrorism, suicide, homicide, illness, or substance abuse, but it refers to hazards holistically;[32] for consistency, the committee also uses the term hazards in this appendix, but it highlights specific concerns about potential losses from acts of terrorism. In the main text of the report, the committee also refers to threats, but does so broadly, to encompass a wide range of tactical and other concerns, including those of terrorism. Whether a hazard or a threat, the condition under consideration can lead to injury, illness, death, or damage, with a particular probability and severity.

Generally speaking, a hazard can be viewed as a condition that might give rise to an event that can be more or less likely (probability) and entail more or less damage (severity). For example, a family might store an ignitable material, such as turpentine, near a gas heater in a basement. This storage arrangement poses a hazard that might lead eventually to a fire, entailing physical injury and property destruction. The odds of the fire could be high or low and the amount of damage could be marginal or catastrophic, depending on whether the basement is cluttered or family members are home. A public safety officer might be interested in helping families to reduce the probability of fire by addressing the conditions under which the fire can occur.

In this report, the committee is interested in identifying conditions under which terrorists can obtain chemicals to produce IEDs and means of reducing the odds that precursor chemicals fall into the wrong hands as fodder for a terrorist episode, which is the ultimate event of concern. To that end, one might consider whether a chemical can be or has been used in an IED and whether and how a terrorist might lay hands on it. Alternatively, the problem could be framed in terms of episode risk (R_E) and precursor risk (R_P). The former is a function of the probability (P_E) and severity (S_E) of the episode (Equation B-1), while the latter is a function of the probability of observing the use of the precursor (P_P) given the episode or concern for the episode and its associated risk (Equation B-2). P_P will be dependent on constituent factors, such as the ease of extracting a precursor chemical from a product or formulation and the attendant personal dangers of handling it.

APPENDIX B

$$R_E = f(P_E, S_E) \tag{B-1}$$

$$R_P = f(P_P|E, R_E) \tag{B-2}$$

The national-security doctrine juxtaposes probability and severity in a risk matrix that can be used to evaluate the risk level of a particular hazard and inform, if not establish, mitigation priorities; the doctrine addresses both severity and probability to distinguish, for example, the hundred-year storm, the tsunami, and seasonal flooding. Therefore, the possibility of a moderately harmful but likely event might generate as much concern as the possibility of a catastrophic but unlikely event. Quantitative data, for example, rates of occurrence, illness, injury, death, or damage, should be used to support the assessment, if available, but are not absolutely necessary.

In the committee's simplified rendering of the standard matrix (Figure B-1), risk levels reside in four quadrants, range from high (red) to low (yellow), and depend equally on probability and severity, but other more-sophisticated formulations are possible. The doctrinal version is asymmetric, provides more categories for probability and severity, and, in some instances, gives slightly more weight to severity than probability.[32,33] Nonlinear permutations are also possible.[237]

The matrix presents an initial basis for considering, if not establishing, priorities across scenarios, chemicals, and policy responses and can be used in at least three ways. One could start a risk analysis from the end of a narrative with a set of scenarios—that is, types of terrorist episodes of particular concern because of their riskiness—and back out to a set of chemicals of interest; one could start the analysis from the beginning with a set of chemicals and track them forward to scenarios, by asking what risks each presents under what conditions, with what likelihood and severity; or one could approach the analysis from both directions and iterate toward convergence.

Preliminary to any such approach, the matrix can be used to establish whether an episode is of particular concern, depending on whether it is sufficiently risky. For example, a scenario involving a larger-scale explosive device, such as a truck bomb, could entail substantially more damage than a scenario involving a smaller-scale, person-borne device, such as a backpack bomb, but could be far less likely; thus, the risk of either scenario might rate concern. If taking the scenario-by-scenario approach, one could identify the chemicals that terrorists can use to produce each type of device and the conditions (e.g., relating to availability and other factors) under which they can obtain them; develop strategies to reduce the odds of terrorists getting access to the chemicals; and, ultimately, lessen the risk of either scenario. By limiting the analysis to a set of scenarios, one can focus attention on a single dimension of precursor risk, namely, the probability of observing the use of the precursor, given concern for the episode (Equation B-2). Figure B-2 redraws the matrix as a continuum without strict

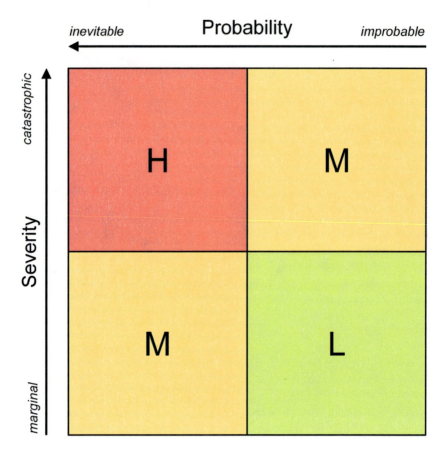

FIGURE B-1 Simplified risk matrix; H: high (red), M: medium (orange); L: low (yellow).

rankings, in which the risks of vehicle- and person-borne episodes are at similar levels and potentially reducible.

However, the matrix, even with elaboration, cannot provide definitive guidance for establishing priorities, in part, because it is silent on the feasibility and costs of mitigation options. If relying only on the matrix, a policy maker might inadvertently ignore low-hanging fruit or suggest dedicating inordinate resources to the impossible; a low risk might have an easy and cheap fix whereas a medium risk might have no feasible or affordable remedy.

Researchers offer a number of criticisms of qualitative tools and methods of analysis, including risk matrices. While acknowledging the popularity of matrices, they warn against a lack of scientific validation and the drawbacks of using them indiscriminately.[243,244] For example, correlations between severity

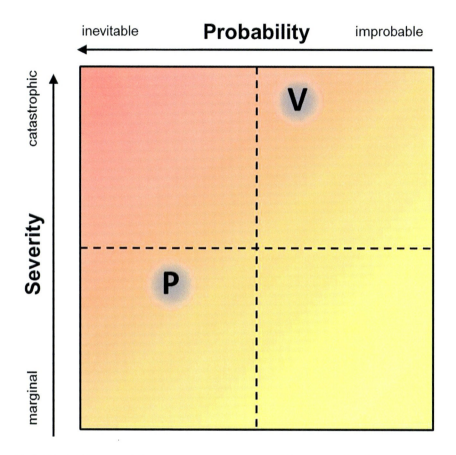

FIGURE B-2 Simplified risk matrix as a continuum with two scenarios; P: episode involving person-borne device; V: episode involving vehicle-borne device.

and probability can invert results. Lundberg and Willis offer remarks on the imperfection of various analytical options:[246] "While some articles have explored the reasons to be cautious about the use of qualitative risk assessment tools there are also reasons to be cautious about the use of quantitative estimates"; the latter point is further explored elsewhere.[245]

The decision to start or stop mitigating risk also requires consideration of risk tolerance and the acceptability of residual risk. The former refers to the level of risk below which a hazard does not warrant any expenditure of resources on mitigation, and the latter refers to the amount of risk that remains after mitigation. If the initial level of risk is low enough to be tolerable absent any mitigation, then the residual risk would equal the initial risk. By implication, risk mitigation need not amount to risk elimination. In some instances, it might be preferable to accept some amount of risk—initial or residual—and develop a response and recovery plan.[247] Such preferences typically arise when the costs of addressing the risk outweigh the anticipated benefit.

The matrix is also silent on causality, but the cause or causes of a hazard can bear directly on policy making. Absent a thorough understanding of the underlying reasons for a hazard, a policy maker might choose a risk mitigation strategy that does not improve conditions or makes them worse.[247,248] The same can be said of attempts to control risk associated with precursor chemicals.

Fortunately, the risk matrix is just one tool for managing risk, and national-security doctrine nests it in a continuous five-step risk-management process (Figure B-3) that addresses these and other concerns, potentially including those pertaining to operational dynamics and behavioral adaptations. A particular hazard might present a low-level risk in one moment and a high-level risk in another and, even if policy makers have a firm grasp on causality, efforts to control risk can entail unintended consequences that give rise to new and different risks. Taking action in one arena might result in displacement and merely push the risk elsewhere. The five-step risk-management process allows—and even requires—deeper, ongoing consideration of risks, controls, and consequences and provides

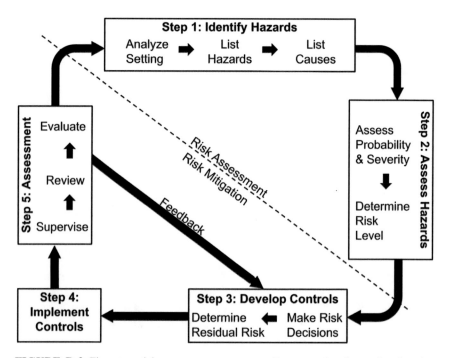

FIGURE B-3 Five-step risk-management process. Recent national-security doctrine includes a much-simplified figure.[247,248]

explicit means to incorporate new information and address changes in circumstances. Whereas the matrix is inherently static, the five-step process, which forms an unending loop, is inherently dynamic. This meets the fourth principle of risk management, "apply risk management cyclically and continuously."[32,33]

In Figure B-3, steps 1 and 2—which include hazard identification and address causality, probability, and severity—constitute risk assessment, and steps 3 through 5 constitute risk mitigation.[247] From the national-security doctrine,[32]

> Steps 3 through 5 are the essential follow-through actions to manage risk effectively. In these steps, leaders balance risk against costs and take appropriate actions to eliminate unnecessary risk and accept residual risk at the appropriate level. During execution, leaders continuously assess the risk to the overall mission and to those involved in the task. Finally, leaders and individuals evaluate the effectiveness of controls and provide lessons learned so that [*they and*] others may benefit from the experience.

Steps 4 and 5 generate experience and feedback, which can either validate the current approach or suggest the need for a change of course or a procedural refinement, whether applying the process in a military or civilian setting.

Appendix C

History of High-Profile Bombing Attacks

Table on following page.

Event (Location)	Main Charge	Mass (lb)	Booster	Initiator
1970-Sterling Hall Bombing (Madison, WI)	ANFO	2,000	Dynamite	Caps
1983-Beirut Barracks Bombing (Beirut, Lebanon)	PETN	20,000	u	u
1983-US Embassy Bombings (Beirut, Lebanon)	ANFO	2,000	u	u
1992-St. Mary Axe Bombing (London, United Kingdom)	CAN/IS	1,000-2,000	S/DC	u
1993-World Trade Center Bombing (New York, NY)	Urea Nitrate	1,200	NG/AN	u
1993-Bishopsgate Bombing (London, United Kingdom)	CAN/IS	2,000-4,000	S/DC	u
1995-Oklahoma City Bombing (Oklahoma City, OK)	AN/NM	5,000	Commercial	Caps
1996-Manchester Shopping Mall (Manchester, United Kingdom)	CAN/IS	1,000-3,000	S/DC	u
1996-South Quay bombing (London, United Kingdom)	CAN/IS	1,000-3,000	S/DC	u
1996-Khobar Towers Bombing (Khobar, Saudi Arabia)	C4	20,000		u
1998-US Embassy Bombings (Tanzania, Kenya)	TNT	2,000		u
1999-Millennial Bomber Interdiction (Port Angeles, WA)	Urea Nitrate	<500	EGDN	HMTD
2000-USS Cole Bombing (Aden, Yemen)	Mil. Exp.	1,000		u
2001-Shoe Bomber (AA Flight 63)	PETN	<1	DC	TATP
2002-Bali Nightclub Bombing (Bali, Indonesia)	KClO$_3$/S/Al	2,000	DC	u
2003-Marriott Hotel Jakarta Bombing (Jakarta, Indonesia)	KClO$_3$/S/Al	100	u	u
2003-British Consulate Bombing (Istanbul, Turkey)	AN/Al	1,000-2,000		u
2003-Casablanca Bombings (Casablanca, Morocco)	TATP/AN	10-20		u
2004-Australian Embassy Attack (Jakarta, Indonesia)	KClO$_3$/S/Al	1,000-2,000	u	u
2004-US Consulate Failed Attack (Karachi, Pakistan)	CHP/Flour	<2,000		Caps
2004-Disrupted Jordanian Attack (Amman, Jordan)	CHP/Cumin	>10,000	NG	Caps
2004-US Embassy Attack (Tashkent, Uzbekistan)	AN/Al	10-20		u
2004-Madrid Train Bombings (Madrid, Spain)	Dynamite	10-20		u
2005-7/7 Underground Bombing (London, United Kingdom)	CHP/Black Pepper	10-20		TATP
2005-7/21 Bombing (London, United Kingdom)	CHP/Flour	10-20		TATP
2006-Operation Overt (London, United Kingdom)	CHP/Tang	<1		HMTD
2006-Disrupted Plot (Ontario, Canada)	ANFO	6,000-7,000	u	u
2007-Disrupted Bomb (Ramstein, Germany)	CHP/Flour	1,000	u	u
2008-US Embassy Attack (Sana'a, Yemen)	TNT	100		u
2009-Underwear Bomber (NWA Flight 253)	PETN	<1		TATP
2009-Operation Highrise Interdiction (Denver, CO/New York, NY)	CHP/Flour	10		TATP
2010-Printer Bombs (United Kingdom/United Arab Emirates)	PETN	<1		MF
2010-Failed Times Square Plot (New York, NY)	AN/IS/Sawdust	100		Pyrotechnics
2011-Khalid Ali-M Aldawsari Plot (Lubbock, TX)	Picric Acid	10-20		TATP
2011-Oslo Bombing (Oslo, Norway)	ANFO/AN/Al/MB	2,000	Picric Acid	DDNP
2012-Aurora Theater Shooting (Aurora, CO)	BP	10-20		KPMG/Thrm
2013-Boston Marathon Bombings (Boston, MA)	Pyrotechnics	10-20		u
2015-Paris Attacks (Paris, France)	TATP	10-20		u
2016-Brussels Attacks (Brussels, Belgium)	TATP	30-40		u
2016-Ahmad Khan Rahami (New York/New Jersey)	AN ET/BP/HMTD	1-10		u
2017-Concert Bombing (Manchester, United Kingdom)	TATP			u

NOTE: AN: ammonium nitrate, AN/FO: ammonium nitrate / fuel oil, PETN: pentaerythritol tetranitrate, CAN: calcium ammonium nitrate, NM: nitromethane, TATP: triacetone triperoxide, BP: black powder, CHP: concentrated hydrogen peroxide, TNT: trinitrotoluene, MB: micro-balloons, Mil. Exp.: military explosives, u: unknown, NG: nitroglycerine, EGDN: ethylene glycol dinitrate, DDNP: diazodinitrophenol, IS: icing sugar, S/DC: semtex/detcord, MF: mercury fulminate, KPMG: potassium permanganate / glycerin, Thrm: thermite, ET: exploding target mixture.

Appendix D

Group A Chemical Supply Chains

The purpose of this appendix is to provide an overview of the Group A precursor chemicals' supply chains and to highlight the unique aspects of the nodes, modes of transportation, and oversight of each. In addition to federally mandated oversight, the commercial handling and movement of precursor chemicals may be subject to state and local laws or ordinances and corresponding regulations depending on the specific jurisdiction, of which there are thousands in the United States. The committee could not review all of their policies on precursor chemicals and has not indicated them on every node in each supply chain, except in the case of fertilizer grade ammonium nitrate (FGAN). Specific data on volumes shipped and imports/exports is excluded in favor of qualitative descriptors, given the fluid nature of commerce from year to year. The general key for reading the diagrams is provided below, along with a glossary defining the different node designations, and is applicable for all diagrams shown.

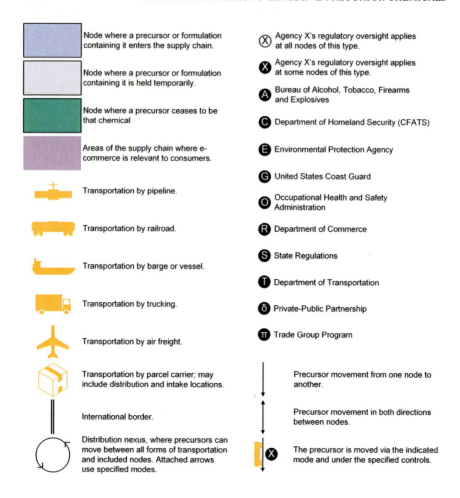

Aerospace – Users that employ the precursor chemicals as aerospace fuels.

Ag. (Agriculture) – End users who employ the precursor chemicals in agricultural production, e.g., farmers, landscapers, and other horticulturalists.

Ag. Retail. (Agricultural Retailer) – Locations that deal solely with agricultural products, likely in bulk forms, and that may provide application services.

Amt. Prd. (Amateur Production) – Production of a precursor chemical by a private individual.

AN Fact. (Ammonium Nitrate Factory) – A factory whose primary product is ammonium nitrate.

Blaster – A commercial operation that uses explosives, e.g., a mine.

Bunker – A temporary storage location used to hold sensitive chemicals.

Chm. Prc. (Chemical Process) – A manufacturer where the chemical is consumed as part of another chemical manufacturing process, e.g., use of nitric acid for polymer synthesis.

Com. Dist. (Commercial Distributor) – A warehouse, wholesaler, or distributor that temporarily holds bulk precursor chemicals or finished products containing the precursor chemicals; may include distributors that only handle chemicals.

Consumer – A person who purchases a precursor chemical or formulation for personal or professional end use, e.g., a hobbyist purchasing racing fuel or a cosmetician purchasing hair bleach.

Ex. Mfr. (Explosives/Propellants Manufacturer) – A manufacturer that incorporates the precursor chemical into commercial explosive materials.

F. Prd. Mfr. (Finished Product Manufacturer) – A manufacturer that incorporates a precursor chemical into a complete consumer product, e.g., ammonium nitrate cold packs.

Factory – A location where a precursor chemical is manufactured from raw materials, e.g., the synthesis of hydrogen peroxide from methane and oxygen.

Hopper – Temporary storage bins used to transfer materials between transportation modes.

Lab. Supl. (Laboratory Supplier) – A company that retails chemicals to researchers, e.g., Sigma Aldrich.

Other Ind. (Other Industries) – Nonchemical process manufacturers that use the precursors, e.g., in metal casting.

Packager – A location where bulk precursor chemical is divided into smaller containers for retail.

Port – A seaport where goods enter or exit the United States, e.g., the Port of Houston. Most ports will use terminals for temporary storage and transloading (an exception being the Port of New Orleans, where some precursors are directly loaded onto barges).

Remed. (Remediation) – A location where chemicals are used for bulk disinfection, e.g., water treatment.

Retailer – An online or physical location where consumers may purchase products consisting of or containing the precursor chemical, e.g., Amazon, Home Depot, and CVS.

ROW (Rest of World) – Other countries that engage in precursor chemical commerce with the United States not otherwise specified on the chart.

ALUMINUM POWDER, PASTE, AND FLAKE

Aluminum powder is commercially manufactured from bulk ingots using a melting and aspiration process to form an aerosol before aggregation (Figure D-1, Factory 1).[56] Paste and flake can subsequently be made from the powder, either in a co-located or other facility (Figure D-1, Factory 2). Paste is made via the addition of solvents to the powder, which is then dried as a thin film to create flake. There is movement of this precursor chemical through ocean ports or across the Canadian border (though it is not made in that country). The majority of aluminum powder, paste, and flake (PPF) is transported via trucks, with railroads being used uncommonly for bulk transport and air freight for sample delivery, noted on Figure D-1 with the letter u. Movement of bulk material through commercial distributors is uncommon, but does occur.

Finished products that include aluminum in an extractable form suitable for IEDs also enter the supply chain and can include paints and coatings, for which precursor-extraction tutorials are available online (these products are not included on the charts once applied to a surface by consumers or other industry).

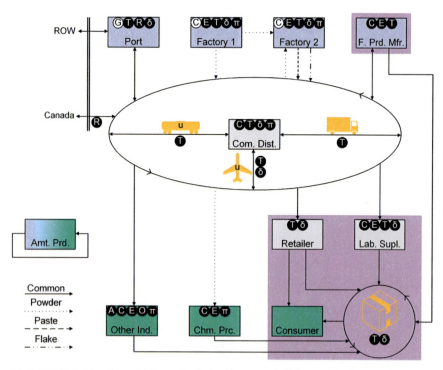

FIGURE D-1 Aluminum PPF supply chain. Common (solid arrow) indicates all types of PPF moving between nodes.

Another relevant finished product is exploding target kits, which contain separate packages of aluminum powder and ammonium nitrate, though no person or data source could tell the committee who is providing the aluminum powder to that finished product manufacturer (communications to exploding target kit manufacturers were unanswered).

Chemical and other industrial end users (see green boxes in Figure D-1) have been known to return excess product to the supply chain through e-commerce transactions, though the specifics of this practice were not provided to the committee. These excess materials are widely available online in pound-scale packages.[249] Private persons are also capable of milling their own powder from bulk aluminum, which could take a week to achieve particles comparable to flake aluminum. The total quantity of aluminum PPF shipped domestically ranges in the tens of millions of pounds per annum.[250]

EPA only covers aluminum "fume or dust" (CAS number 7429-90-5) under EPCRA § 313.[251] This also appears on the TSCA list, but only as "Aluminum," so it is unclear under what conditions TSCA will apply at the factory nodes given the many possible forms of aluminum metal.[252] There is an export restriction on aluminum powder with particle sizes below 200 μm. ATF oversees aluminum PPF at locations where it is mixed to produce ammonal or other commercial explosives and pyrotechnics, as does OSHA. These locations are marked as other industrial end users (Figure D-1, Other Ind.) as those products are beyond the scope of this study.[103] DOT regulation of aluminum powder transported by truck, rail, or plane either as a 4.1 (coated, UN1309) or 4.3 (uncoated, UN1396) material depends on specific product testing and may not apply to a given shipment of powder;[253] if it does qualify as a 4.3 material, a DOT security plan is always required, whereas 4.1 only requires a security plan if part of a desensitized explosive mixture.[113] Private-public partnerships may include C-TPAT at seaports and KSP for aircraft transportation. Trade group programs can apply to both manufacturers and distributors (SOCMA, ACC, and NACD for all but Other Ind., which may include IME).

AMMONIUM NITRATE, FERTILIZER GRADE

FGAN is formed as a downstream process of nitrogen fixation, in which nitric acid and ammonia are ultimately neutralized in an aqueous solution and solidified in a prilling tower (Figure D-2, Factory).[58,150,254] The majority of FGAN produced in the United States is transported in bulk form, with only a minority being packaged, either onsite or at an external packager or distributor. Finished products of note may include cold packs or exploding targets, though the committee was unable to determine the source or precise grade of the AN in those formulations, and the FGAN manufacturers and trade groups that provided information were clear that they were not sure of the source, as they or their members did not directly sell to those finished product manufacturers.

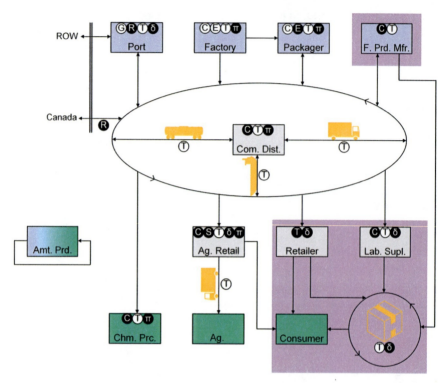

FIGURE D-2 FGAN supply chain.
NOTE: AN can be produced via amateur methods that, although varying in physical makeup, is chemically identical to FGAN/TGAN.

FGAN imports and exports primarily cross the Canadian border in bulk, with the intended destination dependent on the nearest factory or distributor. A smaller amount of material enters and exits through seaports. Bulk shipments typically move via rail, barge, and truck (including transloading), and may be kept at warehouses and distributors en route to end users. The chemical industry is a minority end user, and primarily includes producers of nitrous oxide.[255] Agricultural end users, as defined in Chapter 3, make up the bulk of the end-user base. In many instances, local agricultural retailers are the final intermediary before the fertilizer is applied to the fields, which is a service many of them provide, in which case they maintain custody from the retail location to soil. Other agricultural end users, including landscapers and other horticulturalists, may also purchase FGAN from agricultural retailers. Other retailers are unlikely to carry neat FGAN fertilizer, but instead sell packaged blends with other fertilizers; all of these are also available online. Various research-grade AN products are available from laboratory suppliers, though at significantly higher cost than from other types of suppliers.[256]

There are export restrictions on products containing 15% or more AN. EPA lists AN under TSCA, and under EPCRA § 313 as category code N511, "Nitrate Salts (water dissociable)"; the latter regulation only applies to aqueous solutions and not the solid prills, and is therefore excluded from the majority of the diagram.[251] CFATS applies to those nodes containing more than 2,000 lb of material containing ≥ 33% FGAN.[88] DOT classifies AN as a 5.1 oxidizer and requires a security plan if more than 3,000 kg is transported.[113,253] Some fertilizer blends may be transported as a class 9 miscellaneous HM, provided they meet the requirements of Special Provision 132, which either contains not more than 70% ammonium nitrate, and not more than 0.4% total combustible, organic material calculated as carbon or contains not more than 45% ammonium nitrate and unrestricted combustible material, and therefore would not require a security plan. Trade group programs include programs sponsored by both manufacturers (e.g., ACC) and distributors (e.g., NACD and ResponsibleAg). As defined in Chapter 3, FGAN is further regulated in several states at the agricultural retailer node. KSP may apply to all nodes involved with air shipping, and all seaports include C-TPAT and MTSA.

AMMONIUM NITRATE, TECHNICAL GRADE

Technical grade ammonium nitrate (TGAN) is produced using the same general process as FGAN, the only difference being the increased porosity of the TGAN prills (Figure D-3, Factory).[57] These grades are separated for the purposes of this report due to the different user bases and the segregated supply chains of the two materials. There are scenarios in which a batch of TGAN is returned to the factory due to physical changes during shipment that made the product unusable (indicated by a double-headed arrow).

Like FGAN, only a limited amount of TGAN is bagged before distribution (1.0–1.5 ton packages) or during distribution (smaller bags). Imports from overseas into the United States are possible, but uncommon, with the majority of imports and exports occurring across the Canadian border, depending on proximity to manufacturers. The same finished product manufacturers buying FGAN (e.g., cold packs or exploding targets) might be buying TGAN, though the primary TGAN manufacturers and trade associations could not provide information on where those finished product manufacturers are obtaining their supplies.

Transport primarily occurs in bulk via rail, truck, and barge, with intermodal transloading between shipping modes and temporary storage occurring in railside hoppers. The majority of TGAN is manufactured by or shipped directly to explosives manufacturers or blasters that mix and use it onsite. A small amount is used in chemical processes. TGAN is available at online retailers as finished products (which are frequently bought together with other precursor chemicals),[257] or in small packages of neat product.[258] Exploding target kits are available at physical locations and online directly from the manufacturer.[259,260]

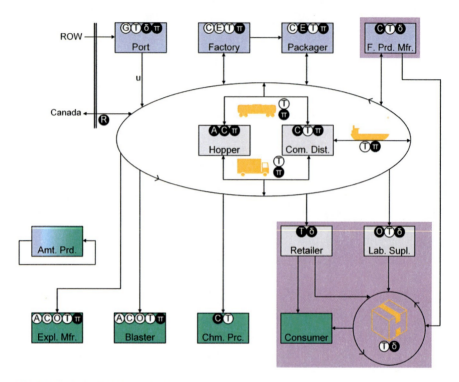

FIGURE D-3 TGAN supply chain.
NOTE: AN can be produced via amateur methods that, although varying in physical makeup, is chemically identical to FGAN/TGAN.

CFATS and EPA do not distinguish between FGAN and TGAN, so the regulations previously described for FGAN apply. While export restrictions do exist for TGAN, the current lack of exports via ocean shipping precludes its inclusion on the port node. The primary trade group program throughout the TGAN supply chain is that of IME, which sets standards for storage and shipping; other programs may include those of chemical manufacturing and distributing associations. ATF regulations apply at locations where explosives are manufactured from TGAN or at locations that mix components onsite (e.g., a mine or job site); ATF also may oversee storage locations, depending on proximity to DOT class 1 explosives. OSHA will specifically apply controls at these locations as well.[103] DOT classifies neat TGAN as a 5.1 oxidizer and requires security plans for shipments in excess of 3,000 kg.[113,253] All mixtures with significant combustibles are listed as class 1 explosives and are beyond the scope of this study. KSP may apply to all nodes involved with air shipping, and all seaports include C-TPAT and MTSA.

CALCIUM AMMONIUM NITRATE

There are currently no domestic producers of calcium ammonium nitrate (CAN), with all domestic quantities being sourced from other countries (Figure D-4).[44] Bulk material typically arrives through seaports (mostly with intermediate terminals existing before entry into the transportation nexus, except for New Orleans, where barges are directly filled from ocean vessels), with some overland transaction from Canada (though they import it first as well). About 5% of the total CAN volume is packaged. Finished products on this diagram include packaged fertilizer blends and cold packs.[261] The same transportation modes and other nodes supply a primarily agricultural end-user base. Finished products are available online and at brick-and-mortar retail locations.

CAN would be subject to export restrictions if it were shipped to other countries. The EPA regulations described for FGAN also apply to CAN. Unlike the other formulations, CAN is not regulated by DOT, as the CAN within the supply chain does not meet the AN concentration threshold of >80% given in Special

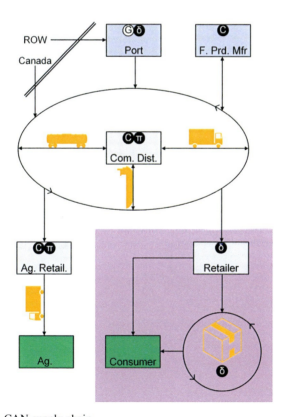

FIGURE D-4 CAN supply chain.

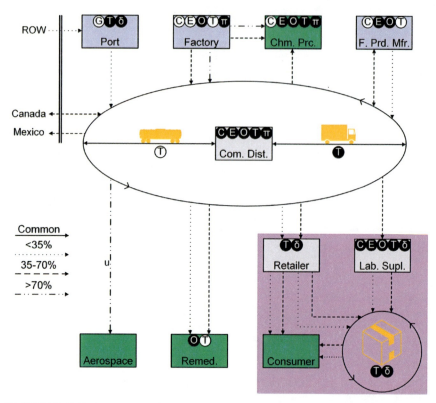

FIGURE D-5 Hydrogen peroxide supply chain. Common (solid arrow) indicates that any concentration may move along that route.

Provision 150 Part b. ResponsibleAg participation may happen at agricultural retailers, and NACD's program may apply to distributors. KSP may apply to all nodes involved with air shipping, and all seaports include C-TPAT and MTSA.

HYDROGEN PEROXIDE

Hydrogen peroxide production usually occurs in close proximity to natural gas sources, which provide the hydrogen feedstock used in the anthraquinone process (Figure D-5, Factory).[55] Some of this production is co-located with other chemical processes (e.g., pulp and paper bleaching), or the hydrogen peroxide product ships primarily to nodes where those processes occur.[148,255] Imports of hydrogen peroxide from overseas are unlikely given the difficulty of transporting highly corrosive, high concentration hydrogen peroxide or the cost of transporting dilute solutions made of mostly water; any imports will probably be finished products that include hydrogen peroxide within their formulations. There is significant overland transaction across the Canadian and Mexican borders; Canada both imports and exports based on factory proximity to the end user, while Mexico only imports from the United States.

Bulk shipments primarily occur in the 50–70% concentration range, with only a small amount of >70% hydrogen peroxide supplying some military and aerospace end users. Finished products may include health and personal care formulations, which are diluted from concentrated bulk, or further purified hydrogen peroxide for specialty applications (e.g., electronics manufacturing). Some distributors dilute as well. During bulk transportation, containers are typically sealed, barcoded, and cross-referenced with paperwork at the shipper and receiver. Locations where hydrogen peroxide is used for soil remediation and water treatment are labeled Remed. in Figure D-5. Some distributors dilute hydrogen peroxide.

Physical retailers do not typically carry products with ≥35% hydrogen peroxide concentration, with pool and spa suppliers[262] using about 27% and personal products being <15% in most cases;[263] online retailers carry these products as well. Lab suppliers typically sell around the 30% range, though concentrations as high as at least 60% are available as well.[264] On some websites, there are formulations available that claim to be 35% and are marketed as part of pseudoscientific health practices (these may also be available at some physical retail locations, so a direct line from retailer to consumer was included as well).[265] Domestic usage of hydrogen peroxide is on the hundreds to thousands of kilotons per annum scale.[266,267]

CFATS applies to locations meeting the weight (400 lb) and concentration (>35%) thresholds;[88] however, remediation nodes are specifically exempted.[89] OHSA applies at locations where 7,500 lb or more of ≥52% hydrogen peroxide is present.[104] EPA uses the same 52% threshold under EPCRA § 302 and § 304;[251] TSCA also applies to relevant production nodes.[252] All hydrogen peroxide formulations >8% are considered by DOT to be class 5.1 oxidizers, requiring a security plan if transported in quantities exceeding 3,000 L;[113,253] additional requirements are provided for solutions >40% in Special Provision 12, and for mixtures with peroxyacetic acid in Special Provision 145. It was reported to the committee that aerospace nodes may be under controls stipulated by the Department of Defense, but the committee was unable to confirm the applicable regulations and left that node blank on the diagram. Chemical manufacturers and distributors may participate in relevant trade association programs. KSP may apply to all nodes involved with air shipping, and all seaports include C-TPAT and MTSA.

NITRIC ACID

Nitric acid production is typically co-located with another process that uses nitric acid for a separate synthesis. This can either be wholly contained, with no nitric acid entering the external commercial supply chain, or, as is the case for AN producers (Figure D-6, AN Fact.), excess product is sold to other end users.[254] If

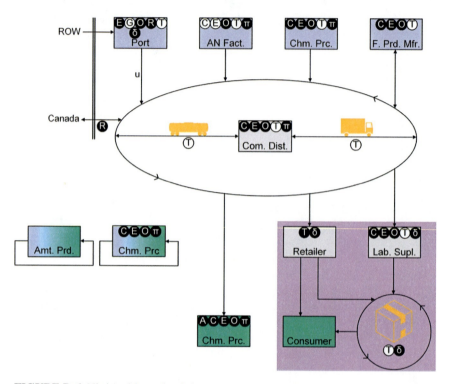

FIGURE D-6 Nitric acid supply chain.

sold, nitric acid moves in a bulk form to the locations of other chemical processes (e.g., equipment maintenance or polymer manufacturing) or to finished product manufacturers and commercial distributors.[255] Bulk shipments to the locations of chemical processes are not commonly routed through distributors, and if that does happen the distributor would not take possession. As with AN and hydrogen peroxide, import and export with Canada occurs, dependent on specific proximities. Imports from overseas, while possible, are uncommon and most likely take the form of finished products. Finished products mainly include cleaning supplies and etchants, which are available at physical and online retail locations.[268] Lab suppliers are known to provide fuming nitric acid that is ≥99% concentration.[269] Online tutorials allow amateur production of nitric acid.

Nitric acid is export restricted for ≥20% formulations.[117] ATF regulatory oversight only occurs at those locations where nitric acid is used to create explosives (Figure D-6, Chm. Prc.). Like other theft or diversion hazards, nitric acid is regulated by CFATS above 400 lb at ≥68% concentration (the water–nitric acid azeotrope, above which it becomes fuming nitric acid).[88] EPA lists nitric acid under TSCA and has two entries for nitric acid: (1) under EPCRA

APPENDIX D   167

(§§ 302, 304, and 313) and (2) under the Comprehensive Environmental Response, Compensation, and Liability Act (CERCLA) at concentration ≥80%. USCG lists nitric acid as a uniquely hazardous material at concentrations greater than 70%, which dictates chemical-specific handling procedures at facilities under USCG control.[95] OSHA oversight is limited to facilities that use ≥94.5% fuming nitric acid at 500 lb or more total quantity.[104] All forms of nitric acid are considered by DOT to be class 8 corrosives, with all mixtures ≥20% requiring a security plan if the quantity shipped is >3,000 L.[113,253] Chemical manufacturers and distributors may participate in relevant trade association programs, including IME for explosives manufacturers. KSP may apply to all nodes involved with air shipping, and all seaports include C-TPAT and MTSA.

NITROMETHANE

Domestic nitromethane production is limited to one facility (Angus Chemical Company), which nitrates propane to form nitroalkanes (see Figure D-7).[59] This facility also produces nitromethane-furfural blends for agricultural customers. Trucks are used exclusively to transport barrels of nitromethane directly from the

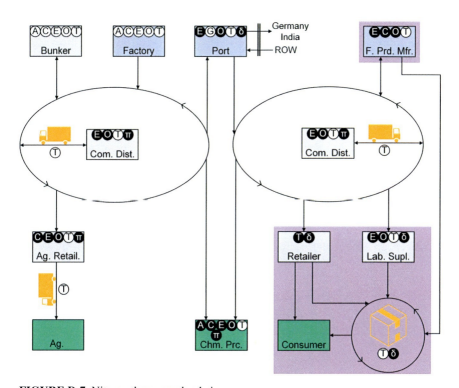

FIGURE D-7 Nitromethane supply chain.

factory to large agricultural distributors, locations of chemical processes, or export locations. Some inventory is stored in an offsite bunker while waiting for shipment. The agricultural distributor delivers smaller portions of the blend to an applicator, who maintains custody until it is applied on a farm. All other end users are supplied by imports from overseas, the bulk being reportedly from China. Finished products include other solvent blends, which feed industrial users, and racing fuels and fuel additives used by hobbyists.[255] Nitromethane is available as these products online and at physical supply stores and at high purity from laboratory suppliers.[270-272]

ATF regulates nitromethane once it is included in specified formulations; though Angus does not make these formulations, Angus representatives claim that they are subject to ATF oversight. CFATS covers all facilities with 400 lb of nitromethane regardless of concentration.[88] EPA regulates nitromethane under § 313 of EPCRA and TSCA.[251,252] Oversight from OSHA only happens once a threshold of 2,500 lb onsite is reached.[104] DOT considers nitromethane a class 3 flammable liquid, which requires a security plan if transported quantities exceed 3,000 L.[113,253] Chemical manufacturers and distributors may participate in relevant trade association programs, including that of IME for explosives manufacturers and ResponsibleAg for agricultural nodes. KSP may apply to all nodes involved with air shipping, and all seaports include C-TPAT and MTSA.

POTASSIUM CHLORATE

The committee was unable to locate an industrial representative willing to speak on potassium chlorate, so the following diagram (see Figure D-8) and information were sourced from other available documentation.[255] Major finished products include matches and specialty agricultural goods. Chemical processes may include making explosives and propellants. Based on DOC data, there appear to be exports to Canada, as well as imports and exports with overseas suppliers. In addition to finished products, neat potassium chlorate is available online both from lab suppliers and retailers.[273,274] Online tutorials demonstrate the capacity for amateur production.

ATF provides oversight of potassium chlorate where it is used to make explosives and propellants, as does OSHA. Potassium chlorate is considered a theft and diversion hazard under CFATS with a 400 lb threshold.[88] Under DOT, potassium chlorate is a class 5.1 oxidizer, which requires a security plan if transported quantities exceed 3,000 kg.[113,253] Chemical manufacturers and distributors may participate in relevant trade association programs. KSP may apply to all nodes involved with air shipping, and all seaports include C-TPAT and MTSA.

POTASSIUM PERCHLORATE

Potassium perchlorate is only manufactured at one domestic location, American Pacific Corporation, which manufactures it as a secondary product of its

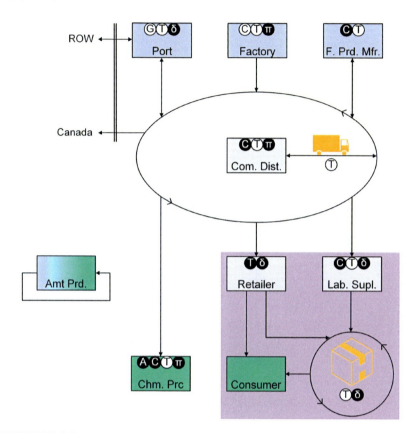

FIGURE D-8 Potassium chlorate supply chain.

ammonium perchlorate process (see Figure D-9).[275] It exclusively supplies the automotive industry, which uses it to make airbags (a finished product). All other shipments originate overseas. Other finished products include small amounts of potassium perchlorate for treating thyroid diseases and for oxygen generators.[255] Other industries are primarily pyrotechnic manufacturers, which use potassium perchlorate instead of chlorates.[276] High-purity potassium perchlorate is available online from various sources in addition to lab suppliers.[277,278]

Potassium perchlorate is regulated by ATF only at those locations where it is used to make explosives or pyrotechnics, or indirectly at the manufacturing location that also produces ammonium perchlorate. CFATS considers it a theft and diversion hazard with a threshold of 400 lb.[88] Under DOT, potassium perchlorate is a class 5.1 oxidizer, which requires a security plan if transported quantities exceed 3,000 kg.[113,253] Chemical manufacturers and distributors may participate in relevant trade association programs. KSP may apply to all nodes involved with air shipping, and all seaports include C-TPAT and MTSA.

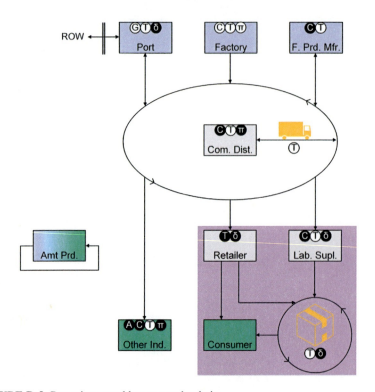

FIGURE D-9 Potassium perchlorate supply chain.

SODIUM CHLORATE

Like hydrogen peroxide, sodium chlorate is primarily manufactured domestically for use in the pulp and paper bleaching process, which may be co-located (see Figure D-10).[55] Some of the sodium chlorate produced is packaged in 1 ton sacks for further distribution. Other chemical processes include metal finishing, refining, making detergents, and explosives manufacturing. There is transaction with Canada dependent on the proximity to the manufacturer, though exports overseas are uncommon. Sodium chlorate can be shipped as either a solid powder or an aqueous solution, though this distinction is not shown on the diagram. Finished products may include herbicides, airbags, and oxygen-generating devices.[255] Most water-treatment facilities (Remed.) that use sodium chlorate install their own generators, which produce what they need onsite, and do not connect with the rest of the supply chain. In addition to finished products, neat sodium chlorate is available online from both lab suppliers and retailers.[279,280] Online tutorials demonstrate the capacity for amateur production. Canada and the United States account for more than half of global sodium chlorate production, on the megaton per annum scale.[281]

APPENDIX D

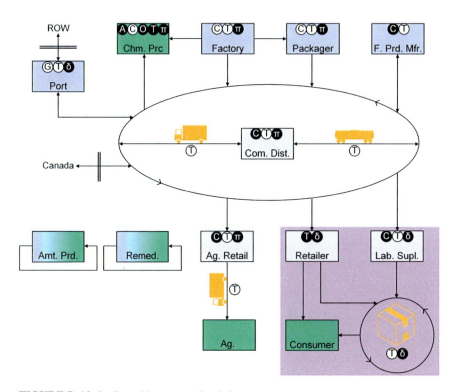

FIGURE D-10 Sodium chlorate supply chain.

Sodium chlorate is regulated by ATF and OSHA only at those locations where it is used to make explosives or pyrotechnics. CFATS considers it a theft and diversion hazard with a threshold of 400 lb.[88] Under DOT, sodium chlorate is a class 5.1 oxidizer, which requires a security plan if transported quantities exceed 3,000 kg.[113,253] USCG lists sodium chlorate as a highly hazardous substance for aqueous solutions of concentration <50% and stipulates additional handling procedures.[95] Chemical manufacturers and distributors may participate in relevant trade association programs. KSP may apply to all nodes involved with air shipping, and all seaports include C-TPAT and MTSA.

UREA AMMONIUM NITRATE SOLUTION

Urea ammonium nitrate (UAN) solution production is typically co-located with AN factories (see Figure D-11), where AN and urea are mixed in an aqueous environment to make UAN.[254] There is some trade between Canada and the United States depending on the nearest manufacturer; the United States also exports overseas. Distribution can happen though truck, rail, barge, or pipeline as

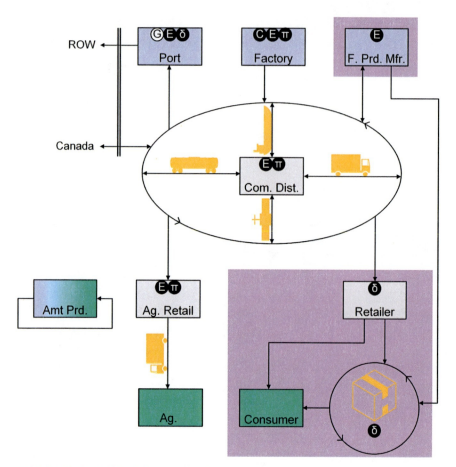

FIGURE D-11 UAN solution supply chain.

UAN solution is a liquid product. It moves from the factories through increasingly smaller distributors until it reaches the agricultural retailer, which may provide application services to local farmers and thus maintain custody of the product to the point of end use. There is at least one supplier of UAN-containing fertilizer products online, though it is unknown from where the contents are sourced.[282]

EPA and USCG both cover UAN under their definitions of aqueous AN.[95,251] Chemical manufacturers and distributors may participate in relevant trade association programs, including ResponsibleAg. All seaports include C-TPAT and MTSA.

Appendix E

International Questions

Thank you for agreeing to meet with us to discuss your approaches to controlling precursor chemicals. We would like to have a conversation about the following questions.

Regulation:
- Can you provide the general history, terms, and objectives of the provisions in the 2013 regulation and background on the decision to update Append II?
- How have the regulatory provisions affected the EU MS and how is the regulation being enforced?
- How does the EU (or EU MS) assess the effectiveness of the regulation?
- How are chemicals secured within industry? How are theft and losses regulated, reported, and investigated by the EU or EU MS?
- How much flexibility do EU MS have in establishing measures, e.g., restricting, banning, or requiring license to purchase for specific chemicals?

Voluntary programs and best practices:
- While developing the 2013 legislation, did you identify any voluntary programs (e.g., training, information sharing, or other private-sector or public-private programs) or best practices that affected the decision to create certain regulatory requirements or decide against others?
- Have any new voluntary efforts emerged since 2013?
- How does the EU (or EU MS) assess the effectiveness of voluntary efforts?

Policy questions:
- What has been your experience with balancing stakeholder concerns in policy making and working toward harmonization across the EU MS?
- How do you engage with stakeholders to obtain feedback on implementation?
- What is the timeline for bringing EU MS into compliance with the regulation?

On behalf of the Committee on Reducing the Threat of IED Attacks by Restricting Access to Chemical Explosive Precursors, we thank you for your assistance with these questions. In addition, if there are any publicly available data or citable documents, relating to the policy making process, implementation, voluntary efforts, assessment, stakeholder engagement, or supply chains that we would be able to use in our report, we would be grateful if you would send them to Camly Tran.

Appendix F

Training Materials

The EU's Standing Committee on Precursors (SCP) issued a guidance document that provides guidance to the authorities of EU MS and to retailers regarding the implementation of the requirements contained in the EU's Regulation 98/2013. The SCP guidelines also offer advice on good practice to retailers.

Four partner countries developed retailer-oriented guidance materials for distribution across the EU MS through a project cofounded by the Prevention of and Fight against Crime program of the Directorate General for Migration and Home Affairs.

Additionally, a template leaflet and a template poster in various languages were developed to provide further practical guidance on how to identify suspicious transactions and on what steps to undertake in case of suspicion (see Figures F-1 and F-2).[283]

Lastly, the SCP developed a document for companies and individuals who sell reportable products via internet markets to members of the public. The document provides advice on marketplace sales of explosives precursors and poisons.

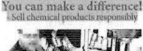

FIGURE F-1 Template leaflet available for distribution across EU MS.

APPENDIX F

SECURITY OF SALES — HIGH RISK CHEMICALS   Revision date: _____

You can make a difference!
- Sell chemical products responsibly

1. Be aware of the products concerned
The following products have been identified as containing chemicals that can be used for malicious purposes:

_____
_____
_____
_____
_____
_____
_____

2. Look out for suspicious behaviour
You should look out for suspicious behaviour in relation to the products concerned. Indicators of suspicious behaviour may include when a customer:
- Appears nervous, avoids communication, or is not a regular type of customer
- Attempts to purchase an unusual amount of a product or unusual combinations of products
- Is not familiar with the regular use(s) of the product(s), nor with the handling instructions
- Is not willing to share what he/she plans to use the product(s) for
- Refuses alternative products or products with a lower (but for the proposed use sufficient) concentration
- Insists on paying cash, especially large amounts
- Is unwilling to provide identity or home address details if requested
- Requests packaging or delivery methods that deviate from what would be ordinary, advised, or expected

3. Report your observations
If you are suspicious of a transaction or attempted transaction, report it to [INSERT: Contact Point] without undue delay.

Try to record as much detail as possible regarding the customer and transaction, such as:
- Height, body type, hair style and colour, facial hair
- Tattoos, piercings, scars, glasses and/or any other distinguishing features
- Registration, make, and model of any vehicle
- Time of purchase, products and amounts involved

Keep any receipts, ID details and CCTV records; any documentation handled by the customer should be preserved for fingerprinting.

CALL [INSERT: Contact Point phone number] AND REPORT

FIGURE F-2 Sample poster available for distribution across EU MS.

Appendix G

Methods and Limitations of Regulatory Assessment

United States federal agencies must satisfy specific requirements for analyzing proposed major or economically significant regulatory actions, such as those expected to have an annual effect on the economy of $100 million or more.[231,236] As required under Executive Order 12866 and other executive branch authorities,[43,284] OMB sets out a hierarchy of methods:[231]

- *benefit-cost analysis* (BCA), in the ideal, monetizes all the benefits and costs of a proposed regulatory action to arrive at a net balance;
- *cost-effectiveness analysis* monetizes the cost of meeting a policy objective, most typically stated as a numerical target, and can be used when costs are known and important categories of benefit can be quantified, but not monetized; and
- *breakeven* or *threshold analysis* asks how much value society must place on the benefits or costs of an action for the benefits to outweigh the costs, and can be used when benefits or costs cannot be monetized or quantified.

As noted below, DHS has employed breakeven analysis in much of its recent analyses of major regulatory actions to address terrorism threats.

BCA, cost-effectiveness analysis, and breakeven analysis have gained acceptance in state and local policy making and among nongovernmental entities, but all three methods have limitations. They do not, for example, address concerns about equity, the distribution of benefits and costs, and the possibility that some members of society might gain while others would not. OMB states, "Regulatory analysis should provide a separate description of distributional effects (i.e., how both benefits and costs are distributed among sub-populations of particular concern) so that decision makers can properly consider them along with the effects on economic efficiency."[231] Moreover, for regulations intended to confer benefits under circumstances of extreme uncertainty, the analytical requirements pose special challenges not limited to data deficiencies.

GUIDANCE ON REGULATORY ASSESSMENT

OMB specifies that a good regulatory analysis, regardless of method, should include "(1) a statement of the need for the proposed action, (2) an examination of alternative approaches, and (3) an evaluation of the benefits and costs—quantitative and qualitative—of the proposed action and the main alternatives identified by the analysis."[231] Before recommending federal regulatory action, an agency must demonstrate that the proposed action is necessary. Questions to consider include the following:

- Is the action intended to address a significant market failure—a situation in which the market for goods or services, alone, cannot provide a satisfactory outcome—or to meet some other compelling public need?
- Could other types of actions—federal or not—be as, or more, effective than the action?
- Could the compelling public need, identified above, be met through state or local regulation, voluntary programs (e.g., programs sponsored by trade associations), public-private partnerships, insurance, or tort liability?

OMB further specifies that a prospective evaluation of benefits and costs should

- explain how the proposed requirements are linked to the expected benefits;
- identify a baseline, typically no action, against which to measure the effects of the proposed regulatory action, and draw comparisons to alternatives; and
- identify and capture the expected undesirable side effects and ancillary benefits of the proposed regulatory action and the alternatives.

OMB also provides the policy community with context, warnings, and caveats. It explains that "the motivation for the analysis is to (1) learn whether the benefits of an action are likely to justify the costs or (2) discover which of various possible alternatives would be the most cost-effective,"[231] but it notes that efficiency might not be the only or overriding public policy objective. OMB also warns that good regulatory analysis is not strictly formulaic; rather, it requires competent professional judgment, transparency, and well-documented supporting evidence, including a discussion of the sensitivity of the results to embedded assumptions. OMB asks agencies to quantify anticipated benefits and costs to the extent possible, but acknowledges that quantification, particularly monetization, may not be feasible across the board. When it is not possible to monetize benefits or costs, an agency may calculate them in terms of physical units; when calculations of physical units are also impossible, an agency may, instead, proceed qualitatively. In particular, OMB[231] recognizes that,

> It will not always be possible to express in monetary units all of the important benefits and costs. . . . If the non-quantified benefits and costs are likely to be important, you should carry out a "threshold" analysis to evaluate their significance. Threshold or "break-even" analysis answers the question, "How small could the value of the non-quantified benefits be (or how large would the value of the non-quantified costs need to be) before the rule would yield zero net benefits?"

None of the methods specified by OMB—BCA, cost-effectiveness analysis, or breakeven analysis—inherently addresses OMB's concerns for either the links between regulatory actions and anticipated benefits or the possibility of unintended consequences and ancillary benefits, nor do any deal directly with matters of equity. The section that follows provides an overview of some of the challenges of assessing control strategies, particularly their benefits.

ANALYTICAL CHALLENGES

Assessing the benefits of actions to prevent or reduce losses from terrorist attacks (e.g., estimates of potential reductions in deaths and injuries, property damage, or various social effects) under extremely uncertain conditions poses substantial challenges.[236,237,285,286] Such an analysis would require an explicit characterization of the underlying risk of a terrorist attack and an understanding of the ways in which the regulations might affect that risk, which would, in turn, present analytical obstacles. In addition, a monetary estimate of benefits might involve evaluating society's willingness to pay to avoid terrorism-related damages.[287,288] Farrow cites specific challenges of interdependencies among defenses or targets; choices among damage- and probability-reducing strategies; the avoidance of unintended consequences in complex systems; ambiguity about the likelihood of events and consequences; the existence of intelligent and adaptive adversaries; and the path-dependent and irreversible nature of some defenses and attack consequences.[289]

Two challenges stand out in this context: (1) a lack of data both for typifying scenarios, even those with a historical basis, and for projecting the nature and likelihood of future incidents; and (2) an inability to anticipate how terrorists will adapt to changes in the security environment.[236,245,290,291] Relevant, reliable, and accurate data are scarce—or unavailable—because terrorist attacks in the United States and comparable settings are infrequent and highly variable and because sensitive information about attacks requires special handling or protection. Moreover, a change in policy that affects the security environment might lead to displacement; in particular, a terrorist could choose a different method of attack or redirect his or her efforts to other settings.[292,293] If, for example, a terrorist finds that is has become too difficult to obtain precursor chemicals to construct IEDs in one location, he or she might choose to work with a vehicle, gun, or knife or to operate elsewhere. The scarcity of data compounds the problem of characterizing potential adaptations and outcomes.

Some recent regulatory assessments have used break-even analysis to circumvent the challenges of benefits estimation. For example, for the proposed Ammonium Nitrate Security Program, DHS compared the anticipated costs of the program to the costs—to people and property—of the bombing of the Murrah Building in Oklahoma City and then calculated the frequency with which similar events must happen to offset the cost of the security program. Members of the policy community have recognized DHS's adoption of break-even analysis as a step forward in the evolution of analysis of terrorism security policy, but have called for further advancement.[237] In a DHS-sponsored workshop held at the RAND Corporation in 2011, the agency itself sought to identify novel approaches to valuing benefits and, potentially, to move beyond break-even analysis.[236] For example, DHS staff presented a fault tree approach that can be used to map existing security programs and policy mechanisms and to estimate event failure probabilities, such as the failure of existing efforts to stop the entry of terrorists into the United States. In developing these models, expert judgment is essential. Quantitative judgments—including quantitative statements of a model's uncertainty—can be combined in these risk models to develop distributions (or ranges) of both baseline risks and changes in risks with the adoption of additional programs and policy mechanisms.

Cost assessment tends to be less fraught because it presents greater certainty. Generally speaking, regulatory agencies propose specific measures that entail known—or knowable—administrative requirements and operational burdens that can be estimated with established accounting and engineering methods. Nevertheless, some elements of cost assessment, such as those that depend on evaluations of businesses' or consumers' behavior or involve unintended consequences, can present uncertainties and pose analytical challenges.

Appendix H

Examples of Retail-Level Control Strategies and Other Measures or Activities

Option	Supplemental Measures	Benefits	Costs[a]	Risks	Precedent
Purchase requires evidence of commercial status, with the right to refuse sales.	• Outreach • Training of retailers to request/verify licenses and identify suspicious behavior • Reporting of suspicious behavior, fraud, theft, or loss • Other documentation (e.g., electronic record keeping for transactions and/or data analytics) • Audits and inspections, including mystery shopping	• Capability to impede all non-commercial acquisitions through legitimate channels (+++) • Awareness of chemicals, concerns, and implementation mechanisms and requirements (++)[b] • Capability to track and correlate suspicious activity, etc., ex ante; investigate incidents, ex post (+) • Capability to provide feedback on implementation (++) • Better visibility of retail transactions for improvements in policy (+) • Non-economic social benefits (+++)	• Public-sector expenditures on administration, outreach, training, data intake and analysis, and audits and inspections (+) • Public-sector expenditures on law enforcement (+)[c] • Private-sector expenditures on administration, outreach, training, reporting, documentation, and audits and inspections (+) • Forgone sales, use, and associated surplus (+++)[d] • Additional transactions time (+) • Non-economic social costs (+++)	• Non-compliance, inadvertent or intentional • Institutional amnesia and/or employee turnover[e] • Outright circumvention, e.g., via diversion or document falsification • Unintentional knowledge transfer • Displacement • Over implementation • Commercial disruption • Discriminatory profiling, potentially offset by training and auditing	• EU/UK regulations • Canadian regulations • Some United States regulations on pesticides • Existing domestic outreach programs that address training and reporting • Existing trade group programs that address training and reporting

Purchase requires either evidence of commercial status or license, with a quantity restriction on licensed purchases and the right to refuse sales.	• Outreach • Training of retailers to request/verify licenses and identify suspicious behavior • Reporting of suspicious behavior, fraud theft or loss • Other documentation (e.g., electronic record keeping for transactions and/or data analytics) • Audits and inspections, including mystery shopping	• Capability to impede illegitimate non-commercial acquisitions through legitimate channels (++) • Awareness of chemicals, concerns, and implementation mechanisms and requirements (++)[b] • Capability to track and correlate suspicious activity, etc, ex ante; investigate incidents, ex post (+++) • Capability to provide feedback on implementation (++) • Vetting for license could draw attention to bad actors (n/a) • Better visibility of retail transactions for improvements in policy (+++) • Non-economic social benefits (++)	• Public-sector expenditures on licenses administration, outreach, training, data intake and analysis, and audits and inspections (+++)[f] • Public-sector expenditures on law enforcement (++)[e] • Private-sector expenditures on licenses, administration, outreach, training, documentation, and audits and inspections (+++) • Forgone sales, use, and associated surplus (++)[d] • Additional transactions time (++) • Non-economic social costs (++)	• Non-compliance, inadvertent or intentional • Institutional amnesia and/or employee turnover[c] • Outright circumvention, e.g., via diversion, improper licensing, or document falsification • Unintentional knowledge transfer • Displacement • Over implementation • Commercial disruption • Discriminatory profiling, potentially offset by training and auditing	• EU/UK regulations • Canadian regulations • Existing state regulations in the United States • Existing domestic outreach programs that address training and reporting • Existing trade group programs that address training and reporting

Option	Supplemental Measures	Benefits	Costs[a]	Risks	Precedent
Purchase requires either evidence of commercial status or signature and government-issued identification, with a quantity restriction on registry purchases and the right to refuse sales.	• Outreach • Training of retailers to request signature and ID at point of sale and identify suspicious behavior • Reporting of suspicious behavior, fraud, theft, or loss • Other documentation (e.g., electronic record keeping for transactions and/or data analytics) • Audits and inspections, including mystery shopping	• Capability to deter and reduce illegitimate non-commercial acquisitions through legitimate channels (+) • Awareness of chemicals, concerns, and implementation mechanisms and requirements (++)[b] • Capability to track and correlate suspicious activity, etc., ex ante; investigate incidents, ex post (++) • Capability to provide feedback on implementation (++) • Better visibility of retail transactions for improvements in policy (+) • Non-economic social benefits (+)	• Public-sector expenditures on administration, outreach, training, data intake and analysis, and audits and inspections (++) • Public-sector expenditures on law enforcement (+++)[c] • Private-sector expenditures on administration, outreach, training, reporting, documentation, and audits and inspections, implementation (++) • Forgone sales, use, and associated surplus (+)[d] • Additional transaction time (+++) • Non-economic social costs (+)	• Non-compliance, inadvertent or intentional • Institutional amnesia and/or employee turnover[e] • Outright circumvention, e.g., via diversion or document falsification • Unintentional knowledge transfer • Displacement • Over implementation • Commercial disruption • Discriminatory profiling, potentially offset by training and auditing	• EU/UK regulations • Canadian regulations • Some United States regulations on pesticides • Existing state regulations in the United States • Existing domestic outreach programs that address training and reporting • Existing trade group programs that address training and reporting

Business as usual with outreach.	• Awareness of chemicals and concerns • Capability to deter and reduce illegitimate non-commercial acquisitions through legitimate channels • Better visibility of retail transactions for improvements in policy • Non-economic social benefits	• Public-sector and/or private-sector expenditures on administration and implementation, including time and materials	• Non-compliance, inadvertent or intentional[g] • Institutional amnesia and/or employee turnover[e] • Unintentional knowledge transfer • Displacement • Over implementation • Discriminatory profiling	• EU/UK regulations • Existing domestic outreach programs that address training and reporting • Existing trade group programs that address training and reporting
Business as usual with training and reporting on suspicious behavior, theft, and loss.	• Awareness of chemicals, concerns, and implementation mechanisms and requirements[b] • Capability to deter and reduce illegitimate non-commercial acquisitions through legitimate channels • Capability to track and correlate suspicious activity, etc., ex ante; investigate incidents, ex post • Better visibility of retail transactions for improvements in policy • Non-economic social benefits	• Public-sector and/or private-sector expenditures on administration and implementation, including time and materials • Public-sector expenditures on data intake and analysis[b] • Public-sector expenditures on law-enforcement • Non-economic social costs	• Non-compliance, inadvertent or intentional[g] • Institutional amnesia and/or employee turnover[e] • Unintentional knowledge transfer • Displacement • Over implementation • Discriminatory profiling, potentially offset by training	• EU/UK regulations • Existing domestic outreach programs that address training and reporting • Existing trade group programs that address training and reporting • Existing commercial practices that are targeted toward other areas

Option	Supplemental Measures	Benefits	Costs[a]	Risks	Precedent
Business as usual with other documentation (e.g., electronic record keeping for transactions and/or data analytics).		• Awareness of chemicals, concerns, and implementation mechanisms and requirements[b] • Capability to deter and reduce illegitimate non-commercial acquisitions through legitimate channels • Capability to track and correlate suspicious activity, etc., ex ante; investigate incidents, ex post • Better visibility of retail transactions for improvements in policy • Non-economic social benefits	• Public-sector and/or private-sector expenditures on administration and implementation, including time, materials, and capital outlays, for record keeping • Public-sector expenditures on data intake and analysis[h] • Public-sector expenditures on law-enforcement costs • Non-economic social costs	• Non-compliance, inadvertent or intentional[g] • Unintentional knowledge transfer[e] • Displacement	• Existing commercial practices that are targeted toward other areas
Business as usual with cautionary labeling.		• Awareness of chemicals and concerns • Product identification • Non-economic social benefits	• Public-sector and/or private-sector expenditures on administration and implementation, including time and materials • Non-economic social costs	• Non-compliance, inadvertent or intentional[g] • Unintentional knowledge transfer	• EU/UK regulations • Existing commercial practices that are targeted toward other areas

| Business as usual with auditing and mystery shopping, partnered with training, reporting, or other measures. | • Awareness of chemicals, concerns, and implementation mechanisms and requirements[b]
• Capability to provide feedback on implementation
• Non-economic social benefits | • Public-sector and/or private-sector expenditures on administration and implementation, including time and materials
• Non-economic social costs | • Non-compliance, inadvertent or intentional[g]
• Unintentional knowledge transfer | • EU/UK regulations |

NOTE: The "+", "++" and "+++" designations represent the relative rankings of the strategy package by type of benefit or cost. A "+++" implies that an option is most beneficial or costly for a particular type of benefit or cost and a "+" implies that it is least beneficial or costly. If all three strategy packages rank equally they obtain a "++". Vetting was unranked (n/a, not applicable) because it pertains only to licensing. a) Expenditures include time and materials and capital outlays, e.g., for record-keeping systems. b) For a voluntary measure, this benefit might be limited to awareness of chemicals, concerns, and implementation mechanisms, absent requirements. c) If retailers report fraud, suspicious behavior, or additional theft and loss, then law-enforcement agencies would incur additional costs. d) Framed as an opportunity cost, this would amount to the difference in the value of alternatives, if available, to sellers and buyers along the supply chain and at retail. e) Institutional amnesia and employee turnover might also be treated as a subset of or reason for noncompliance. f) If licenses are fee-based, then public expenditures on licensing and related administration would be limited to those exceeding the fee. g) Noncompliance as a contractual or regulatory matter could occur under a voluntary or mandatory program, respectively. h) Assignment of any cost to the public sector envisions that some aspects of data intake and analysis, additional to that covered under implementation, would involve sensitive information or require access to restricted information that would be handled by a government agency or contractor.

Appendix I

Committee Member and Staff Biographies

COMMITTEE

Victoria A. Greenfield (*Chair*) is a visiting scholar in the Department of Criminology, Law and Society at George Mason University. She specializes in national security and international social and economic issues, including transnational crime, with a particular interest in drug production and trafficking. In addition, she advises federal agencies and others on strategic planning, performance evaluation, and program management. Some of her latest publications explore means of assessing the harms of criminal activities and the seriousness of crime, the supply chain for doping products, mechanisms for reducing opium-poppy cultivation in Afghanistan, and benefits and costs of terrorism regulation and border security. Dr. Greenfield has also written extensively on international trade, globalization, and defense economics. In recent years, she has held the positions of senior economist, RAND Corporation (currently adjunct), and Admiral Crowe Chair of the Economics of the Defense Industrial Base, U.S. Naval Academy. She also served on the National Academies' Committee on Estimating Costs to the Department of Justice of Increased Border Enforcement. Previously, she was the senior economist for international trade and agriculture, President's Council of Economic Advisers, White House; chief international economist, Bureau of Economic and Business Affairs, U.S. Department of State; and principal analyst, U.S. Congressional Budget Office. She holds a BS in agricultural economics (Cornell University) and a PhD in agricultural and resource economics (University of California–Berkeley).

Robert G. Best is a DOD civilian with the Defense Threat Reduction Agency's Joint Improvised-Threat Defeat Organization (JIDO), hired as a highly qualified expert to design and implement a joint-forces counter-IED signatures program. In this role, he applies his comprehensive knowledge of spectral phenomenology and evolving processes, expert knowledge of state-of-the-art MASINT component and system development, and skill in effective program management techniques. He leads the multiagency IPT and coordinates the development of the C2S2 system and the critical signatures measurement program. In addition, he serves as an interface between the intelligence community, R&D community, and JIEDDO to help identify, evaluate, and coordinate the technical development of new MASINT sensors and/or exploitation hardware/software systems. He works closely with the JIEDDO Intel Division and COIC analysts to help them develop an operational understanding of the observables and exploitable signatures suitable for network attack. Dr. Best has more than 30 years of experience in remote sensing, digital image processing, and spatial information systems. His career has been primarily focused on applications in counter-terrorism, counter-CC&D, counter-narcotics, counter-IED, and radiological emergency response. He is a recognized subject matter expert in MASINT signatures phenomenology, sensor systems, digital image processing, and exploitation. He has a proven history in developing concepts into viable programs. He has designed, integrated, and deployed MASINT systems into operational environments for the intelligence community and U.S. military services. Dr. Best has BS degrees from South Dakota State University in wildlife and fisheries science and general chemistry (both 1974), an MS in wildlife and fisheries science from South Dakota State University (1979), and a PhD in environmental engineering from the University of Wisconsin–Madison (1988).

Leo E. Bradley is an independent consultant and owner of LE Bradley Consulting LLC. He is currently consulting with the Energetics Research Group at Johns Hopkins University Whiting School of Engineering as well as several private companies. He speaks and writes frequently on counter-IED, combating weapons of mass destruction, explosive ordnance disposal (EOD), and explosives safety. He has extensive knowledge on EOD gained during his 29 years as part of the U.S. Army. He served as the commander of the 184th Ordnance Battalion from 2004 to 2006 out of Fort Gillem, Georgia, and when deployed to Iraq. He then served as the chief of the army's EOD Division from 2006 to 2008 and worked with NATO groups to establish common policies and standards relating to EOD and for countering IEDs. From 2008 to 2010 he was assigned to the Office of the Secretary of Defense as chief of EOD and Humanitarian Mine Action. In 2011 and 2012 he led the Combined Joint Task Force Paladin (counter-IED) in Afghanistan, which was charged with defeating IEDs, with marked success. From 2010 to 2012 he served as the commander of the 71st Ordnance Group out of Fort Carson, Colorado, and worked with local, state, and federal law enforcement in the United States for safe and effective EOD. In 2012 he was assigned to the DOD Explosives Safety Board and worked to standardize explosive disposal procedures. Mr. Bradley was a cofounder of the group Return to Adventure in 2011, which is a charity with the goal of assisting the recovery and rehabilitation of wounded EOD and law enforcement bomb squad personnel through outdoor activities and adventure sports. He completed a BS in science at Pennsylvania State University in 1985.

John C. Brulia worked in the commercial explosives industry from 1973 to 2016, most recently at the Austin Powder Company (now retired). Prior to his commercial explosives career, He served as a demolitionist with the U.S. Army Special Forces from 1967 to 1969 in Asia, including duty in Special Operations in Vietnam. After his discharge, he continued his military service as a demolitionist in various National Guard and Reserve Special Forces units until 1977. From 1973 to 2002, Mr. Brulia worked for the Atlas Powder Company (and its successors, ICI Explosives and Orica USA) in multiple roles, including one as a manager of technical development in the area of emulsion explosives, another leading a team of scientists and engineers as a manager of bulk explosives, and another as a vice president of a wholly owned subsidiary responsible for the purchase, storage, transportation, sale, and use of explosives. He worked internationally for 8 years with Orica prior to his retirement in 2002. Subsequently, he served 3 years as the president of Maurer and Scott Inc., a privately held company involved in commercial explosives distribution and specialized blasting services for mining operations. At Austin Powder from 2005 to 2016, he was the director of Safety and Compliance, working with regulatory agencies and responsible for developing security plans, including those for several explosives manufacturing facilities. While at Austin Powder, he also worked as a member representative with the Institute of Makers of Explosives (IME), where he served a term as the chairman of its Security Committee. Throughout his career, Mr. Brulia presented technical papers and wrote multiple articles and handbook chapters on explosives safety, security, and regulation. Since his retirement in February, he has worked with the Bureau of Alcohol, Tobacco, Firearms and Explosives (ATF) to provide advanced explosives training, volunteers with the International Society of Explosives Engineers (ISEE), and continues to serve in his 28th year on the New York State Blasters Examination Board. He has held a blasting license in New York for more than 40 years. He obtained a BA in government in 1972 at Pennsylvania State University.

Carrie L. Castille is an agriculture and natural resources consultant in Louisiana. She served as the former associate commissioner for the Louisiana Department of Agriculture and Forestry for public policy and as senior adviser to Commissioner Mike Strain. She assumed the role at the department after a 10-year tenure as an assistant professor at the Louisiana State University (LSU) Agricultural Center. She holds a BS in engineering from the University of Louisiana at Lafayette, an MS in environmental toxicology from LSU, and a PhD in renewable natural resources with a minor in political science from LSU. She is a fellow of the Food Systems Leadership Institute and serves as the chair-elect on the National Agriculture Research Extension Education and Economics Board appointed by USDA Secretary Vilsack. She also served on the National Council on Environmental Policy and Technology convened by EPA Secretary Gina McCarthy. She created the very successful Louisiana Master Farmer Program while at LSU and continues to work closely with agriculture and forestry producers on national and state policy issues, including agricultural economic development, food safety, environment and natural resources, international trade, and agriculture labor.

David G. Delaney has been a senior fellow at the Center for Health and Homeland Security since 2016. Previously, he taught at Indiana University while serving as the deputy director and senior fellow of the Center for Applied Cybersecurity Research. Mr. Delaney served as a deputy associate general counsel at the Department of Homeland Security where he advised and coordinated efforts related to security and law enforcement. After law school he was a law clerk to Judge James E. Baker of the U.S. Court of Appeals for the Armed Forces. Previously, he had been a military police officer from 1994 to1999 at positions in the United States and Europe. David earned a BS from the U.S. Military Academy at West Point, an MA in law and diplomacy from Tufts University, and a JD from Boston College.

Arthur G. Fraas joined Resources for the Future (RFF) as a visiting fellow in April 2009 after a distinguished career in senior positions within the federal government. In 2008, he retired after 21 years as chief of the Natural Resources, Energy and Agriculture Branch, Office of Information and Regulatory Affairs, Office of Management and Budget (OMB). Much of his work has examined the federal regulatory process, with a particular focus on the impact of environmental regulations. At RFF, Dr. Fraas works on a variety of issues related to energy and the environment, including projects looking at the tradeoffs between using biomass in transportation and in electronic applications, the treatment of uncertainty in regulatory analysis of major rules, and the potential regulation of greenhouse gases under the Clean Air Act. Before joining OMB, he was a senior economist at the Council on Wage and Price Stability, a staff member of the Senate Judiciary Subcommittee on Antitrust and Monopoly, an assistant professor of economics at the U.S. Naval Academy, and a staff economist with the Federal Reserve System. He graduated from Cornell University in 1965 with a bachelor's degree in engineering physics, and earned his doctorate in economics from the University of California at Berkeley in 1972.

William J. Hurley has been with the Institute for Defense Analyses (IDA) since 1985 and is currently an adjunct research staff member in the Joint Advanced Warfighting Division. Prior to joining IDA, Dr. Hurley was with the Center for

Naval Analyses. His research has addressed a wide variety of defense issues with emphases on joint forces, analytical methods, concept development, and advanced technologies. He has directed or coauthored more than 50 studies sponsored principally by offices within the DOD. Recent focal points have included urban operations, irregular warfare, and countering IEDs. Dr. Hurley has participated on several study panels addressing these areas and organized by the Defense Science Board, the Academies, and NATO. In addition to his research responsibilities, he was the associate program director and then program director of the Defense Science Study Group (DSSG) from 1991 to 1998. The DSSG is an ongoing program of education and study that introduces outstanding young professors of science and engineering to current issues of national security and military systems and organizations. The program is sponsored by the Defense Advanced Research Projects Agency. Dr. Hurley's academic background is in theoretical physics. He received a BS in physics from Boston College (1965) and a PhD in physics from the University of Rochester (1971), and held research positions at Syracuse University and the University of Texas.

Karmen N. Lappo graduated from the University of Michigan with BS degrees in mechanical engineering and materials science and engineering. She then completed a Masters of Science in Mechanical Engineering at the University of Texas in Austin. Ms. Lappo has been on staff at Sandia National Laboratories since late 2003 working with a variety of energetic materials technical groups. She has worked on explosive component design and production, hydrocode modeling, field test design, explosive tool development and characterization, explosive mixture characterization, and energetic threats evaluation. In 2010, Ms. Lappo accepted and commenced a 2- year assignment with the Department of Homeland Security Science and Technology Directorate as an energetic materials subject matter expert. She served on requirements review panels and several interagency working groups focused on ammonium nitrate safety, security vulnerabilities, and its malicious use in explosive mixtures/devices. After returning to Sandia in 2012 she has led test series to evaluate ammonium nitrate mixtures detonability in support the Department of Homeland Security proposed ruling on ammonium nitrate. The test results were used to advise DHS policy decisions regarding ammonium nitrate security. Ms. Lappo continues to lead high fidelity explosive test activities to include technical evaluations focused on helping industry and government better understand the security vulnerabilities of ammonium nitrate based products.

Becky D. Olinger is detailed to the National Nuclear Security Administration's Office of Counter Terrorism and Counter Proliferation as a science adviser. Previously, she was a program manager at the Los Alamos National Laboratory Global Security Office, where she managed efforts in nuclear security. From 1991 to 1998 Dr. Olinger was a research assistant at the New Mexico Institute of Mining and Technology Energetic Materials Research and Testing Center studying HMEs and explosive materials. She worked as a senior scientist and program manager from 2004 to 2006, working on projects to counter IEDs for law enforcement and the DOD. In 2006 she started working at Los Alamos as a nuclear weapons engineer and project leader on National Nuclear Safety Administration projects. From 2007 to 2009 she was the DE-1 deputy group leader and program manager, supervising a team of researchers studying problems related to counter-terrorism and HMEs. Dr. Olinger received a PhD in physical chemistry from the University of New Mexico–Albuquerque in 2005, where she used infrared spectroscopy techniques to study a variety of materials, including explosives.

Jimmie C. Oxley is a professor of chemistry at the University of Rhode Island (URI), codirector of the Forensic Science Partnership of URI, and team-lead for and former director of the DHS Center of Excellence for Explosives Detection, Mitigation, and Response. Dr. Oxley's lab specializes in the study of energetic materials including explosives, propellants, and pyrotechnics. She has organized and chaired numerous symposia and short courses for government and industrial laboratories on topics ranging from hazards analysis to bomb threats, has testified before Congress on the topic of explosive materials and detection, and has authored numerous papers on energetic materials. She is an elected fellow of the North American Thermal Analysis Society and a reviewer for FBI, NSF, and the National Academies. She has served on six NRC panels including the Military Science Board advising the army on chemical weapon destruction (1998–1999); the Chemistry Board advising ATF and Congress, on the Committee on Marking, Rendering Inert, and Licensing of Explosive Material (1997–1998); the National Materials Advisory Board (NMAB) advising the FAA, on the Committee on Commercial Aviation Security (1995–1998); the Manufacturing Board's Committee on Advanced Energetic Materials (2001–2002); the Naval Studies Board's Determining Basic Research Needs to Interrupt the Improvised Explosive Device Delivery Chain (2005–2008); and the Army Research Lab's Armor and Armaments panel (2009–2011). Jimmie earned a PhD in chemistry from the University of British Columbia in 1983.

Kevin F. Smith is a career supply chain practitioner and president and CEO of Sustainable Supply Chain Consulting, which was founded in 2009 to provide advice and guidance to large-scale supply chains and related businesses concerning strategic planning and organizational development. Mr. Smith served for 8 years as senior vice president of Supply Chain & Logistics for CVS Pharmacy, the retail arm of CVS Caremark, where his role was to facilitate changes in the overall supply chain. He has been a longtime board member for the Council of Supply Chain Management Professionals (CSCMP), Special Advisor to Supply Chain 50, and contributor to the Retail Industry Leaders Association (RILA). Additionally, he is vice chair of the Distribution Business Management Association (DBMA)

Supply Chain Leaders in Action Executive Committee. Mr. Smith is a graduate from the University of Massachusetts with a Bachelor of Arts degree in English.

Kirk Yeager has been a senior forensic scientist at the Federal Bureau of Investigation (FBI) since 2010. He completed his BS in chemistry in 1987 at Lafayette College and obtained his PhD in inorganic chemistry at Cornell University in 1993. From 1994 to 1995 he worked as a postdoctoral research scientist at the New Mexico Institute of Mining and Technology and researched the hazards and terrorist potentials of energetic materials. Starting in 1995 he held an adjunct faculty appointment at the New Mexico Institute of Mining and Technology and taught several courses on energetic chemistry in addition to conducting research on the production of explosive materials and their safe utilization. In 2000 Dr. Yeager joined FBI as a physical scientist and forensic examiner with the explosives unit. In this capacity he conducted examinations of crime scenes involving bombing and arson, and developed training procedures for law enforcement bomb squads. As a senior scientist with the explosives unit he served as an international expert in explosive and hazardous devices and managed research programs dedicated to creating characterization techniques for emerging HME materials. In 2010 Dr. Yeager was promoted to the executive service, holding positions as FBI's chief explosives scientist and as a senior-level science adviser. He has provided expert testimony in criminal cases involving explosives and technical presentations to government agencies, law enforcement organizations, and Congress. He has published many articles, research reports, and book chapters on the topic of explosives and has been recognized with several awards, including the FBI Director's Award twice. He also serves on the International Association of Bomb Technicians and Investigators advisory committee.

STAFF

Camly Tran joined the Board on Chemical Sciences and Technology at the National Academies of Sciences, Engineering, and Medicine in 2014 as a postdoctoral fellow after receiving her PhD in chemistry from Brown University and is currently a program officer. During her time at Brown, she received various honors including the Elaine Chase Award for Leadership and Service, the American Chemical Society Global Research Exchanges Education Training Program, and the Rhode Island NASA grant. Dr. Tran completed the workshop summaries and reports of *Mesoscale Chemistry* (2015) and *The Changing Landscape of Hydrocarbon Feedstocks for Chemical Production* (2016). She has also supported the consensus studies *Spills of Diluted Bitumen from Pipelines: A Comparative Study of Environmental Fate, Effects, and Response* (2016); *Chemical Laboratory Safety and Security: A Guide to Developing Standard Operating Procedures* (2016); *Effective Chemistry Communication in Informal Environments* (2016); and *Communicating Chemistry: A Framework for Sharing Science: A Practical*

Evidence-Based Guide (2016). She is currently supporting activities on chemistries of the microbiomes, chemical weapons, and precursor chemicals for IEDs.

Teresa Fryberger joined the National Academies in 2013 to serve as director of the Board on Chemical Sciences and Technology. She is an accomplished research program and policy manager in the chemical, environmental, and energy sciences, with experience at NASA, the White House Office of Science and Technology Policy, the Department of Energy, and the Brookhaven and Pacific Northwest National Laboratories. Earlier in her career, Dr. Fryberger was an associate editor at *Science* and a National Research Council postdoctoral fellow at the National Institute for Science and Technology. Teresa earned her PhD in physical chemistry from Northwestern University and her BS in chemistry from the University of Oklahoma.

Samuel Goodman joined the Board on Chemical Sciences and Technology as a postdoctoral fellow in 2016. He graduated with a BS in chemical engineering from the University of Wisconsin–Madison in 2012 and continued his graduate studies at the University of Colorado–Boulder. He completed his PhD in chemical engineering in 2016, with a research focus on the biological impacts of semiconducting nanomaterials.

Jarrett Nguyen is a Senior Program Assistant for the Board on Chemical Sciences and Technology. He began working for the National Academies in October 2016. He graduated from James Madison University in 2015 with a BS in geology and environmental science and a minor in geographic science.